HANDBOOK OF ELECTRONIC MATERIALS
Volume 5

HANDBOOK OF ELECTRONIC MATERIALS

Compiled by:
ELECTRONIC PROPERTIES INFORMATION CENTER
Hughes Aircraft Company
Culver City, California

Sponsored by:
AIR FORCE MATERIALS LABORATORY
Air Force Systems Command
Wright Patterson Air Force Base, Ohio

Volume 1:
OPTICAL MATERIALS PROPERTIES, 1971

Volume 2:
III-V SEMICONDUCTING COMPOUNDS, 1971

Volume 3:
SILICON NITRIDE FOR MICROELECTRONIC APPLICATIONS, PART I:
PREPARATION AND PROPERTIES, 1971

Volume 5:
GROUP IV SEMICONDUCTING MATERIALS, 1971

In preparation:

Volume 4:
NIOBIUM ALLOYS AND COMPOUNDS

Volume 6:
SILICON NITRIDE FOR MICROELECTRONIC APPLICATIONS, PART II: APPLICATIONS

Group IV Semiconducting Materials

M. Neuberger

Electronic Properties Information Center
Hughes Aircraft Company, Culver City, California

IFI/PLENUM · NEW YORK-WASHINGTON-LONDON · 1971

This document has been approved for public release and sale;
its distribution is unlimited. Sponsored by : Air Force Materials
Laboratory, Wright-Patterson Air Force Base, Ohio.

Library of Congress Catalog Card Number 76-147312

ISBN-13: 978-1-4684-7919-5   e-ISBN-13: 978-1-4684-7917-1
DOI: 10.1007/978-1-4684-7917-1

FOREWORD

This report was prepared by Hughes Aircraft Company, Culver City, California under Contract Number F33615-70-C-1348. The work was administered under the direction of the Air Force Materials Laboratory, Air Force Systems Command, Wright-Patterson Air Force Base, Ohio, with Mr. B. Emrich, Project Engineer.

The Electronic Properties Information Center (EPIC) is a designated Information Analysis Center of the Department of Defense, authorized to provide information to the entire DoD community. The purpose of the Center is to provide a highly competent source of information and data on the electronic, optical and magnetic properties of materials of value to the Department of Defense. Its major function is to evaluate, compile and publish the experimental data from the world's unclassified literature concerned with the properties of materials. All materials relevant to the field of electronics are within the scope of EPIC: insulators, semiconductors, metals, super-conductors, ferrites, ferroelectrics, ferromagnetics, electroluminescents, thermionic emitters and optical materials. The Center's scope includes information on over 100 basic properties of materials; information generally regarded as being in the area of devices and/or circuitry is excluded.

Grateful acknowledgement is made for the review and comments by Dr. Victor Rehn of the U.S. Naval Ordnance Test Station at China Lake, California, as well as for review by staff members of the National Bureau of Standards, National Standard Data Reference System.

v

CONTENTS

INTRODUCTION

The Electronic Properties Information Center has developed the Data Table as a compilation of the most reliable information available in the literature, on all properties of a given material. These Data Tables originally served as an introduction to the graphic data compilations published by this Center. These Data Sheets are principally concerned, according to the scope of the Center, with electronic and optical data, but it is believed that data covering the complete property spectrum is of the first importance to every scientist and engineer, whatever his information requirements. For this reason, the Data Table was formulated to include all the physical, mechanical, thermal, electronic, electrical, magnetic and optical properties of the specific material. Increasing requests for this highly selective type of information resulted first in compilation of the III-V Semiconductor Compounds Data Tables and then in these, the Group IV Materials.

The major problem in this type of selective data compilation on a semiconducting material, lies in the material purity. Properties may vary so widely with doping, crystallinity, defects, geometric forms and the other parameters of preparation, that any attempts at comparison normally fail. On this basis, we have consistently attempted to give values derived from experiments on the highest purity single crystals or epitaxial films. At the very least, these data should be reproducible and this gives the data their principal validity. If such values however, are not available, then the next best data are reported, together with material specifications. These latter include the carrier concentration and the dopant. Although the Tables include three elements and one binary compound, the importance of doping is so pre-eminent, that diffusion coefficients and energy levels have been included.

Values for a range of temperatures, wavelengths, frequencies, pressures and field strengths, (both electrical and magnetic), are reported when available. Our primary goal has been not to compress, but to select and present a rounded and fully representative view of the specific material.

This comprehensive review of each material has been made possible by the extensive collection of documents in the EPIC files; to date over 45,000 technical journal articles and Government reports have been acquired by the Center. To compile these Group IV Semiconducting Materials Data Tables, about 3000 articles on Germanium alone, 2250 on Silicon, 400 on Silicon Carbide and 325 on Diamond were evaluated for relevant data.

As far as possible, the arrangement of data has been standardized in a consistent order as follows:

PHYSICAL, MECHANICAL, THERMAL

Property	Unit
Formula	
Molecular Weight	
Density	g/cm^3
Name	
Mineral Name	
Color	
Hardness	Mohs, kg/mm^2
Cleavage	
Symmetry	
Space Group	
Lattice Parameters	$\overset{\circ}{A}$
Melting Point	$^{\circ}C$
Sublimation Temperature	
Specific Heat	cal/gr $^{\circ}K$
Debye Temperature	$^{\circ}K$
Thermal Conductivity	W/cm $^{\circ}K$
Thermal Expansion Coefficient	$10^{-6}/^{\circ}K$

Property	Unit
Elastic Coefficient	
Compliance, s	$cm^2/dyne$
Stiffness or Elastic Modulus, c	$dyne/cm^2$
Shear Strength	kg/cm^2
Young's Modulus	$dynes/cm^2$
Poisson's Ratio	
Sound Velocity	cm/sec
Compressibility (1/Bulk Modulus)	$cm^2/dyne$

ELECTRICAL, ELECTRONIC

Property	Unit
Dielectric Constant	
Static, ε_o	
Optic, ε_∞	
Dissipation Factor, tg δ	
Electrical Resistivity	ohm-cm
Mobility	$cm^2/V\ sec$
Electron, μ_n	
Hole μ_n	
Temperature Coefficient, T^x	
Microwave Emission	
Lifetime, τ	sec.
Cross-section, σ	cm^2
Piezoelectric Coefficients	$C/N,\ C/m^2,\ m/V$
Electromechanical Coupling Coefficient	
Piezoresistance Coefficients	$cm^2/dyne$
Elastoresistance Coefficients	
Effective Mass	m_o
Diffusion Coefficients	cm^2/sec
Energy Levels	eV
Energy Gap	eV
Temperature Coefficient, dE/dT	$eV/°K$
Pressure Coefficient, dE/dP	$eV/kg\ cm^{-2}$
Field Coefficient	
Deformation Potential	eV
Photoelectric Threshold	eV
Work Function	eV
Electron Affinity	eV
Barrier Heights	eV
Phonon Spectra	meV
Seebeck Coefficient	$V/°K$
Nernst-Ettingshausen Coefficient	
Magnetic Susceptibility	cgs
g-Factor	
Superconducting Transition Temperature	$°K$

OPTICAL

Property	Unit
Transmission	%
Absorption Coefficient	cm^{-1}
Refractive Index	
Spectral Emissivity	

Property	Unit
Piezo-optic Coefficient	$cm^2/dyne$
Elasto-optic Coefficient	
Electro-optic Coefficient	
Laser Properties	

Changes in value with temperature and pressure are always given where available. The units have been standardized as far as possible in the cgs system, except for piezoelectric coefficients which, according to usage in this country, are given in Coulombs/Newton; certain mechanical properties data are given in psi.

The most highly valued aspect of this work, is the fact that every individual data point is accompanied by a reference citation. This allows the reader to refer to the original research paper for additional information and offers a representative bibliographic review of these materials. Where two or more documents present the same data values, all are cited. The bibliography which follows each set of tables, is arranged alphabetically by author; more than one document by the same author is distinguished by the letters A, B, C, etc.

These Data Tables begin with a comparative Data Table which lists several key properties of the Group IV materials. These latter are arranged in order of increasing molecular weight, whereas the sections in the main body of the compilation are arranged according to the material name, alphabetically. The comparative table serves not only the convenience of the reader, but indicates as well, the gaps in our knowledge of these materials.

GROUP IV SEMICONDUCTING MATERIALS COMPOSITE DATA TABLE

Material	Diamond	Silicon		Silicon Carbide		Germanium	
Density (g/cm^3)	3.515	2.32831		3.2		5.3243	
Symmetry	cubic	cubic		cubic	hexagonal	cubic	
Lattice Parameters (Å)	3.56679	3.43089		4.3595	a_o= 3.08065 c_o=15.11738	5.65754	
Melting Point (°C)	3827	1420		2830		941	
Thermal Conductivity (W/cm°K)	20	1.4		0.2	4.9	0.586	
Thermal Expansion (10^{-6}/°K)	1	2.44		2.9		5.5	
Dielectric Constant — Static	5.93	11.9		9.72	10.32	16.0	
Dielectric Constant — Optical					6.7		
Electrical Resistivity (ohm-cm)	10^{13}	2×10^5		0.7	1	200	
Mobility (cm^2/V sec)		20°K	300°K			4°K	300°K
Electron	2000	10^5	1880	980	300	10^6	4×10^3
Hole	1500	8×10^4	400		50	5×10^5	2×10^3
Effective Mass (m_o) — Electron	0.25		1.18	0.4	1.5	see Table	
Hole	1.1		0.81		3.5	see Table	
Energy Gap (eV) — Direct	7.4		3.07	6.0		0.889	0.805
Indirect	5.47	1.15	1.1135	2.5	2.994	0.741	0.664
Work Function	6.02		4.6		4.4	4.80	
Refractive Index $_{(Na)}$	2.4186		3.4 (1-10µ)	2.48	2.648(n_o) 2.689(n_e)	5.6	

All data, unless otherwise specified, is given for high purity material at 300°K

PHYSICAL PROPERTIES	SYMBOL	VALUE	UNIT	NOTES	TEMP.(°K)	REFERENCES
Formula		C				
Atomic Weight		12.011				
Density		3.515	g/cm^3			Bochko & Orlov
		3.434		carbonado		
Mineral Name		diamond				
		carbonado		naturally sintered diamond grains		
Types		Ia		nitrogen-platelets in [100]		Evans & Phaal
		Ib		single nitrogen in substitutional sites, yellow		Angress et al.
		IIa		low-nitrogen, low in crystal defects		Clark et al. A, Roberts & Walker
		IIb		p-type, semiconducting, blue, aluminum-doped		Lightowlers & Collins
Color		white, colored to black		transparent		
Hardness		10		Mohs		
Knoop Microhardness	H_{100}	8820±1380	kg/mm^2			Goryunova, p. 65
Cleavage		[110]		perfect		Dana
Symmetry		cubic				Donnay
Space Group		Fd3m Z-2				Donnay
Lattice Parameter	a_o	3.56679	Å			Donnay
	C-C	1.54				Wyckoff
Symmetry		hexagonal		P~130 kbar, T>1000°C		Bundy & Kasper
Space Group		$P6_3/mmc$ Z-4				Bundy & Kasper
Lattice Parameters	a_o	2.52	Å			Bundy & Kasper
	c_o	4.12				
Melting Point		4100	°K	P=125 kbars, diamond-graphite-liquid eutectic		Bundy, Gold
Combustion Point		500	°C	in air		Sokhor & Vitol
Specific Heat		0.00115	cal/mol °K	grains	12.8	Desnoyers & Morrison, Burk & Friedberg
		0.0157			68.6	
		0.1124			120.3	
		0.7761			225.6	
		1.2332			274.1	
		1.2319			275	De Sorbo
		1.4805			300	
		3.242		grains	500	Gmelin, p. 289
		4.660			800	
		5.380			1100	

DIAMOND

PHYSICAL PROPERTIES	SYMBOL	VALUE	UNIT	NOTES	TEMP.($^\circ$K)	REFERENCES
Debye Temperature		2240±5	$^\circ$K	elastic constants	0	McSkimin & Bond
		2219±20		specific heat	0	Desnoyers & Morrison
		2246±15		specific heat	0	Burk & Friedberg
		2010			100	
		1845			200	
		1861		specific heat	300	De Sorbo
Thermal Conductivity		0.1	W/cm $^\circ$K	types, I, IIa, IIb	3	Berman et al., A & B
		100			80	
		20			300	
		100		IIb	100	Dean et al.
Thermal Expansion Coeff.		1.0-1.2	$10^{-6}/^\circ$K	lattice constant meas.	293	Sokhor & Vitol
		3.7		natural and synthetic	>675	
		0.4		gem-quality	173	Thewlis & Davey
		1.0			300	
		2.8			475	
		4.0			875	
		4.8			1175	
Elastic Coefficients						
Stiffness	c_{11}	10.76	10^{12}cm^2/dyne	type I, acoustic meas. at 20-200 MHz	300	McSkimin & Bond
	c_{12}	1.25				
	c_{44}	5.76				
Sound Velocity			Orientation			
Longitudinal		1.85	[111] 10^6cm/sec.		300	Arlt & Schodder
		1.83	[110]	type I, measured at	300	McSkimin & Bond
		1.75	[001]	20-160MHz		
Shear		1.28	[110]			
		1.28	[001]			
Compressibility		0.226	10^{-12}cm^2/dyne		300	McSkimin & Bond

ELECTRICAL PROPERTIES	SYMBOL	VALUE	UNIT	NOTES	TEMP.($^\circ$K)	REFERENCES
Dielectric Constant						
Static	ε_o	5.93		type IIa	300	Gibbs & Hill
		5.87		type I		
		5.46		calc.		Vinsom & Jaros
Pressure Coeff. $1/\varepsilon_o(d\varepsilon/dP)_T$		-1.07	10^{-7}cm^2/kg	type IIa P to 6000 kg/cm^2	300	Gibbs & Hill
		-2.35		type I		
Temperature Coeff. $1/\varepsilon(d\varepsilon/dT)_P$		1.1	$10^{-5}/^\circ$K		325-475	Narasimhan
Electrical Resistivity		10^{13}	ohm-cm	type IIa, high purity,	300	Mykolajewycz et al.
		3×10^{11}		defect-free	500	
		2×10^5		IIb	200	Dean et al.
		100			300	
		<10			1000	

DIAMOND

ELECTRICAL PROPERTIES	SYMBOL	VALUE	UNIT	NOTES	TEMP.(°K)	REFERENCES
Electrical Resistivity		6×10^2		natural IIb	300	Lightowlers & Collins
		2×10^3		Al-doped synthetic		
		40–100		B-doped synthetic		
Mobility						
Electron	μ_n	2000	cm^2/V sec	type I, $n_n = 10^{18} - 10^{19}\ cm^{-3}$	300	Konorova & Shevchenko, Redfield
Hole	μ_p	1500		type I		
Temperature Coeff.	μ_n, μ_p	$\sim T^{-1.5}$			100–300	
Hole	μ_p	1550		type IIb	300	Dean et al.
Temperature Coeff.	μ_p	$\sim T^{-3}$			>400	
Drift Mobility						
Electron	μ_{Dn}	2000		type I, E to 2 kV/cm	293	Konorova & Shevchenko
		8000		E to 500 V/cm	120	
Hole	μ_{Dp}	6000		E to 600 V/cm	293	
Lifetime						
Electron		10^{-2}	μsec	type IIa, electroluminescence	4	Vavilov et al. A
		$10^{-2} - 10^{-5}$		Ia, IIa, electrical meas.	300	Konorova et al.
Hole		9		IIb, photoconductivity	150	Johnson et al.
Piezoresistance Coeff.	π_{11}	-0.43	$10^{-13} cm^2/dyne$	type Ia, optical meas. at 0.583μ	300	Burstein & Smith
	π_{12}	+0.37				
	π_{44}	-0.27				
Effective Mass						
Electron	m_n	0.25	m_o	IIb, electrical meas. $n_p = 5 \times 10^{16}$	300	Wedepohl
Hole	m_p	1.1		IIb, optical meas., $n_p = 10^{16}$	300	Dean et al.
Light Hole	m_{lp}	0.70		IIb, cyclotron resonance meas.	1–4	Rauch A & B
Heavy Hole	m_{hp}	2.12				
Split-off V.B.	m_{so}	1.06				
Electron Density of States	m_{Dn}	0.2		IIb, optical meas.	90–600	Clark et al. B
Energy Levels Dopant	E_a					
Al	0.373		eV	IIb, photoconductivity	4–150	Collins & Lightowlers

ELECTRICAL PROPERTIES	SYMBOL	VALUE	UNIT	NOTES	TEMP.($^\circ$K)	REFERENCES
Energy Levels Dopant	E_a					
B		0.35	eV	IIb, electrical and optical meas. on synthetic crystals	300-800	Bezrukov et al.
		0.19		natural type I diamond, ion implantation	300-1200	Vavilov et al. B & C
N		4.05		IIa, photoconductivity	300	Denham et al.
Neutral Center	N9	5.251, 5.261		no-phonon lines, luminescence meas.	100	Wight & Dean, Crowther & Dean
Ionized Center	N3	2.985		N-A1 center, absorption and luminescence meas.	100	Crowther & Dean
				absorption meas., type I	300	Sobolev et al.
Energy Gap						
Direct $\Gamma_{25}'-\Gamma_{15}$	E_o	7.3	eV	IIa, optical meas.	0	Roberts & Walker
		7.4		IIa, optical meas.	300	Walker & Osantowski
Indirect $\Gamma_{25}-\Delta_1$	E_g	5.470		IIb, optical transmission at 5-6 eV, 90-600°K	295	Dean & Crowther, Clark et al. B
	E_o	7.12 7.02			133 295	
Temperature Coeff.	dE_o/dT dE_g/dT	-6.3 -0.54	10^{-4} eV/$^\circ$K	IIb, optical transmission	90-600	Clark et al. B
Pressure Coeff.						
Uniaxial	dE_g/dP	0.19	10^{-6} eV/kg cm^{-2}	IIa, IIb, absorption meas. P=30,000 kg/cm^2	80	Dean & Crowther
Hydrostatic		0.5		P=45,000 kg/cm^2		
Deformation Potential						
Indirect Gap		-2.3	eV	IIb, absorption and luminescence meas.	80-300	Dean & Crowther
Valence Band	b d	+2.47 +1.63		IIb, uniaxial stress, absorption	80	Crowther et al.
Photoelectric Threshold	Φ	6.02	eV	IIb, electron photoemission		Leivo et al.
Work Function	ϕ	6.02	eV	IIb, electron photoemission		Leivo et al.
Electron Affinity	ψ	0.9		IIb, electron photoemission		Leivo et al.
Phonon Spectra						
Longitudinal Optic	LO	163	meV	Ia, low nitrogen, luminescence meas.	100	Wight & Dean
Transverse Optic	TO	141				
Transverse Acoustic	TA	65				
Longitudinal Acoustic	LA					
Raman Phonon	R	166				
	LO	144		I, IIa, absorption meas. at 2-6μ	300	Hardy & Smith
	TO	158				
	TA	92				
	LA	123				

ELECTRICAL PROPERTIES	SYMBOL	VALUE	UNIT	NOTES	TEMP.(°K)	REFERENCES
Phonon Spectra						
	LO	147.6	meV	Ia, inelastic neutron	1000°C	Peckham
	TO	129		scattering on [100]		
	TA	95.4				
	LA	147.6				
	LO	132		IIb, absorption meas.	300	Clark et al. B
	TO	143		at 5-14μ, T=90-600°K		
	TA	83				
	LA	132		luminescence meas. at		Dean & Jones
	R	167		0.25μ, T=90-320°K		

	Γ	X	L	W	NOTES		REFERENCES
LO		146.9	155.2	146.2	1st and 2nd Raman Spectra		Solin & Ramdas
TO		132.6	149.5	123.8	for IIa, IIb, at 4, 77, and		
TA		100.0	70.1	112.6	300°K		
LA		146.9	124.7	146.2			
R	165.0						

ELECTRICAL PROPERTIES	SYMBOL	VALUE	UNIT	NOTES	TEMP.(°K)	REFERENCES
Seebeck Coeff.		10	mV/°K	IIb	220	Goldsmid et al.
		3.5			300	
		2			400	
		1.2			700	
Magnetic Susceptibility		-0.456	10^{-6} cgs	magnetic meas., I, II	300	Sigamony
		-0.57		x-ray analysis of electron distribution	290	Sirota & Sheleg
g-Factor	g_1	2.0031		Ib, electron spin resonance	300	Klingsporn et al.
	g_2	2.0019				
	g_3	2.0025				
	g	2.003		IIb	300	Bell & Leivo

OPTICAL PROPERTIES	SYMBOL	VALUE	Wavelength (μ)	NOTES	TEMP.(°K)	REFERENCES
Transmission		50 %	0.25-2.5	IIb	300	Johnson et al.
Refractive Index	n	2.4186	0.578	IIa	300	Larsen & Berman, Waxler & Weir
		2.3	1.2	I	300	Philipp & Taft
		3.5	0.177			
Pressure Coeff.	dn/dP	-1×10^{-7}/kg cm^{-2}	0.587	I	300	Waxler & Weir
Temperature Coeff.	dn/dT	1×10^{-5}/°K	0.587	I	300	Waxler & Weir
Density Coeff.	$\rho\,(dn/d\rho)$	+1.24		calc. from photo-elastic constants		Burstein & Smith
		-1.58		from dn/dP		Waxler & Weir
Photoelastic Constants	p_{11}	-0.125			300	Burstein & Smith
	p_{12}	+0.325				
	p_{44}	-0.11				

ANGRESS, J.F. et al. One-Phonon Band-Mode Absorption by Impurity Resonances in Diamond and Silicon. ROYAL SOC. OF LONDON, PROC., v. 308, no. 1492, Dec. 1968. p. 111-124.

ARLT, G. and G.R. SCHODDER. Some Elastic Constants of SiC. ACOUSTICAL SOC. OF AMERICA, J., v. 37, no. 2, Feb. 1965. p. 384-386.

BELL, M.D. and W.J. LEIVO. Electron Spin Resonance in Semiconducting Diamonds. J. OF APPLIED PHYS., v. 38, no. 1, Jan. 1967. p. 337-339.

BERMAN, R. et al. The Thermal Conductivity of Diamond at Low Temperatures. ROYAL SOC. OF LONDON, PROC., SER. A. v. 220, no. 1140, Oct. 22, 1953. p. 171-183. [A]

BERMAN, R. et al. The Thermal Conductivity of Dielectric Crystals: The Effect of Isotopes. ROYAL SOC. OF LONDON, PROC., SER. A, v. 237, no. 1210, Nov. 1956. p. 344-354. [B]

BEZRUKOV, G.N. et al. Some Electrical and Optical Properties of Synthetic Semiconducting Diamonds Doped with Boron. SOVIET PHYS. SEMICONDUCTORS, v. 4, no. 4, Oct. 1970. p. 587-590.

BOCHKO, V.A. and Yu.L. ORLOV. Variation of Density in Varieties of Natural Diamonds. SOVIET PHYS.-DOKLADY, v. 15, no. 3, Sept. 1970. p. 204-207.

BUNDY, F.P. Direct Conversion of Graphite to Diamond in Static Pressure Apparatus. J. OF CHEM. PHYS., v. 38, no. 3, Feb. 1963. p. 631-643.

BUNDY, F.P. and J.S. KASPER. Hexagonal Diamond-A New Form of Carbon. J. OF CHEM. PHYS., v. 46, no. 9, May 1, 1967. p. 3437-3446.

BURK, D.L. and S.A. FRIEDBERG. Atomic Heat of a Diamond from 11 Degrees to 200 Degrees K. PHYS. REV., v. 111, no. 5, Sept. 1, 1958. p. 1275-1282.

BURSTEIN, E. and P.L. SMITH. The Photoelastic Properties of Diamond. PHYS. REV., v. 74, Dec. 1948. p. 1880-1881.

CLARK, C.D. et al. Intrinsic Edge Absorption in Diamond. ROYAL SOC. OF LONDON, PROC., A, v. 277, no. 1370, Feb. 11, 1964. p. 312-329. [B]

CLARK, C.D. et al. The Absorption Spectra of Natural and Irradiated Diamonds. ROYAL SOC. OF LONDON, PROC., SER. A, v. 234, no. 1198, Feb. 21, 1956. p. 363-381. [A]

COLLINS, A.T. and E.C. LIGHTOWLERS. Photothermal Ionization and Phono-Induced Tunneling in the Acceptor Photo-conductivity Spectrum of Semiconducting Diamond. PHYS. REV., v. 171, no. 3, July 15, 1968. p. 843-855.

CROWTHER, P.A. et al. Excitation Spectrum of aluminum Acceptors in Diamond under Uniaxial Stress. PHYS. REV., v. 154, no. 3, Feb. 15, 1967. p. 772-785.

CROWTHER, P.A. and P.J. DEAN. Phonon Inter-actions, Piezo-Optical Properties and the Inter-Relationship of the Nitrogen 3 and Nitrogen 9 Absorption-Emission Systems in Diamond. J. OF PHYS. AND CHEM. OF SOLIDS, v. 28, no. 7, July 1967. p. 1115-1116.

DANA, J.D. System of Mineralogy, 7th Ed. PALACHE, C. et al. v. 1, N.Y., John Wiley & Sons, Inc., 1944, 834 p.

DEAN, P.J. et al. Intrinsic and Extrinsic Recombination Radiation from Natural and Synthetic Aluminum-Doped Diamond. PHYS. REV., v. 140, no. 1A, Oct. 4, 1965. p. A352-A368.

DEAN, P.J. and P.A. CROWTHER. The Effect of Uniaxial and Hydrostatic Pressure on the Absorption Edge Spectrum and the Edge Excitation Spectrum for Visible Luminescence Diamond. In INT. CONF. ON SEMICONDUCTOR PHYS., PROC., 7th, Paris, 1964. v. 4, Ed. by: HULIN, M. N.Y., Acad. Press, 1964. p. 103-112.

DEAN, P.J. and I.H. JONES. Recombination Radiation from Diamond. PHYS. REV., v. 133, no. 6A, Mar. 16, 1964. p. A1698-A1705.

DENHAM, P. et al. Ultraviolet Intrinsic and Extrinsic Photoconductivity of Natural Diamond. PHYS. REV., v. 161, no. 3, Sept. 15, 1967. p. 762-768.

DESNOYERS, J.E. and J.A. MORRISON. The Heat Capacity of Diamond between 12.8 and 227 K. PHIL. MAG., v. 3, no. 25, Jan. 1958. p. 42-49.

DeSORBO, W. Specific Heat of Diamond at Low Temperatures. J. OF CHEM. PHYS., v. 21, no. 5, May 1953. p. 876-880.

DONNAY, J.D.H. (Ed.) Crystal Data. Determinative Tables. 2nd Ed. American Crystallography Assoc., 1963.

EVANS, T. and C. PHAAL. Imperfections in Type I and Type II Diamonds. ROYAL SOC. OF LONDON, PROC., A, v. 270, no. 1343, Dec. 11, 1962. p. 538-551.

GIBBS, D.F. and G.J. HILL. The Variation of the Dielectric Constant of Diamond with Pressure. PHIL. MAG., v. 9, no. 99, Mar. 1964. p. 367-375.

GMELINS HANDBUCH DER ANORGANISCHEN CHEMIE; achte vollig neu bearbeitete Auflage. Carbon. Teil B. Weinheim, Verlag Chemie, GmbH, 1967

PENNSYLVANIA STATE UNIV., UNIVERSITY PK., PA. Coll. of Earth and Mineral Sci. Natural and Synthetic Diamonds and the North American Outlook, by GOLD, D.P., v. 37, no. 5, Feb. 1968. p. 37-48.

GOLDSMID, H.J., et al. The Thermoelectric Power of a Semiconducting Diamond. PHYS. SOC., PROC., v. 73, Mar. 1959. p. 393-398.

GORYUNOVA, N.A. The Chemistry of Diamond-like Semiconductors. Ed. J.D. ANDERSON, The M.I.T. Press, Mass. Inst. of Tech., Cambridge, Mass. 1963, 236 p.

HARDY, J.R. and S.D. SMITH. Two-phonon Infra-red Lattice Absorption in Diamond. PHILOSOPHICAL MAG., v. 6, no. 69, Sept. 1961. p. 1163-1172.

JOHNSON, C., et al. Photoeffects and Related Properties of Semiconducting Diamonds. J. OF PHYS. AND CHEM. OF SOLIDS, v. 25, no. 8, Aug. 1964. p. 827-836.

KLINGSPORN, P.E. et al. Analysis of an Electron Spin Resonance Spectrum in Natural Diamonds. J. OF APPLIED PHYS., v. 41, no. 7, June 1970. p. 2977-2980.

KONOROVA, E.A., et al. Ionization Currents in Diamonds During Irradiation with 500-1000 keV Electrons. SOVIET PHYS.-SOLID STATE, v. 8, no. 1, July 1966. p. 1-5.

KONOROVA, E.A. and S.A. SHEVCHENKO. Investigation of the Carrier Mobility in Diamonds. SOVIET PHYS.-SEMICONDUC-TORS, v. 1, no. 3, Sept. 1967. p. 299-304.

LARSON, E.S.and H. BERMAN. The Microscopic Determination of the Nonopaque Minerals. U.S. Dept. of the Interior Geological Survey, Bull. 848, 2nd Ed. U.S. Govt. Printing Office, Washington, D.C. 1934.

LIGHTOWLERS, E.C. and A.T. COLLINS. Electrical-Transport Measurements on Synthetic Semiconducting Diamond. PHYS. REV., v. 151, no. 2, Nov. 11, 1966. p. 685-688.

OKLAHOMA STATE UNIV. OF AGRICULTURE AND APPLIED SCI., RES. FOUNDATION. STILLWATER. Photoelectric Emission and Surface States of Semiconducting Diamonds. AFCRL-65-489 FR, by LEIVO, W.J. et al. Contract no. AF 19 628-2385. May 28, 1965. 108 p. AD 619 724.

McSKIMIN, H.J. and W.L. BOND. Elastic Moduli of Diamond. PHYS. REV., v. 105, no. 1, Jan. 1957. p. 116-121.

MYKOLAJEWYCZ, R., et al. Some Physical Properties of Nearly Perfect Natural Diamond. APPLIED PHYS. LETTERS, v. 6, no. 11, June 1, 1965. p. 227-228.

NARASIMHAN, P.T. Temperature Dependence of Dielectric Constant of Diamond. PROC. PHYS. SOC. LONDON, v. B68, part 5, May 1955.

PECKHAM, G. The Phonon Disperson Relation for Diamond. SOLID STATE COMMUNICATIONS, v. 5, no. 4, Apr. 1967. p. 311-313.

PHILIPP, H.R. and E.A. TAFT. Optical Properties of Diamond in the Vacuum Ultraviolet. PHYS. REV., v. 127, no. 1, p. 159-161, July 1, 1962.

RAUCH, C.J. Millimeter Cyclotron Resonance Experiments in Diamond. PHYS. REV. LETTERS, v. 7, no. 3, Aug. 1, 1961. p. 83-84. [A]

RAUCH, C.J. Millimeter Cyclotron Resonance in Diamond. In, INTERNAT. CONF. ON THE PHYS. OF SEMICONDUCTORS, PROC. Held at Exeter, July 1962. Ed. by: STICKLAND, A.C. London, Inst. of Phys. and the Phys. Soc., 1962. p. 276-280. [B]

REDFIELD, A.G. Electronic Hall Effect in Diamond. PHYS. REV., v. 94, no. 3, May 1, 1954. p. 526-537.

ROBERTS, R.A. and W.C. WALKER. Optical Study of the Electronic Structure of Diamond. PHYS. REV., v. 161, no. 3, Sept. 1967. p. 730-735.

SIGAMONY, A. Magnetic Susceptibility of Diamond. INDIAN ACAD. OF SCIENCES, PROC., A, v. 19, May 1944. p. 310-314.

SIROTA, N.N. and A.U. SHELEG. Magnetic Suceptibility of Semiconducting Elements of Group IV Determined on the Basis of X-Ray Analysis. SOVIET PHYS.-DOKL., v. 8, no. 9, Mar. 1964. p. 887-889.

SOBOLEV, E.V. et al. Aluminum Impurities in Diamond Absorption Spectra. SOVIET PHYS., SOLID STATE, v. 11, no. 1, July 1969. p. 200-202.

SOKHOR, M.I. and V.D. VITOL. X-Ray Investigation of Thermal Expansion in Synthetic and Natural Diamonds. SOVIET PHYS. CRYSTALLOGRAPHY, v. 11, no. 4, Jan/Feb. 1970. p. 632-633.

SOLIN, S.A. and A.K. RAMDAS. Raman Spectrum of Diamond. PHYS. REV. B, Ser. 3, v. 1, no. 4, Feb. 15, 1970. p. 1687-1698.

THEWLIS, J. AND A.R. DAVEY. Thermal Expansion of Diamond. PHIL. MAG. v.1, no. 5, May 1956. p. 409-414.

VAVILOV, V.S. et al. Recombination Radiation from Diamonds Caused by Electron Excitation. SOVIET PHYS. SOLID STATE, v. 8, no. 5, Nov. 1966. p. 1210-1214. [A]

VAVILOV, V.S. et al. Investigation of the Hall Effect in p-Type Semiconducting Diamond Doped with Boron by the Ion Implantation Method. SOVIET PHYS. SEMICONDUCTORS, v. 4, no. 1, July 1970. p. 12-16. [B]

VAVILOV, V.S. et al. Investigation During Isochronous Multistage Annealing of the Electrical Conductivity of Semiconducting n- and p-Type Diamonds Prepared by the Ion Implantation Method. SOVIET PHYS. SEMICONDUCTORS, v. 4, no. 1, July 1970. p. 6-11. [C]

VINSOME, P.K.W. and M. JAROS. The Microscopic Dielectric Function in Silicon and Diamond. J. OF PHYS. C, v. 3, no. 10, Oct. 1970. p. 2140-2145.

WALKER, W.C. and J. OSANTOWSKI. Ultraviolet Optical Properties of Diamond. PHYS. REV., v. 134, no. 1A, Apr. 6, 1964. p. A153-A157.

WAXLER, R.M. and C.E. WEIR. Effect of Hydrostatic Pressure on the Refractive Indices of Some Solids. NAT. BUREAU OF STANDARDS, J. OF RES., v. 69A, no. 4, July-Aug. 1965. p. 325-333.

WEDEPOHL, P.T. Electrical and Optical Properties of Type IIb Diamonds. PHYS. SOC., PROC., B, v. 70, pt. 2, Feb. 1957. p. 177-185.

WIGHT, D.R. and P.J. DEAN. Extrinsic Recombination Radiation from Natural Diamond, Exciton Luminescence Associated with the N9 Center. PHYS. REV., v. 154, no. 3, Feb. 15, 1967. p. 689-696.

WYCKOFF, R.W.G. Crystal Structures. N.Y. Wiley & Sons, v. 1. p. 114.

PHYSICAL PROPERTIES	SYMBOL	VALUE	UNIT	NOTES	TEMP.($°K$)	REFERENCES
Formula		Ge				
Atomic Weight		72.60				
Density		5.3243	g/cm^3	single crystal	298	Gmelin, p. 83
		5.51		liquid	950°C	Glazov et al. A
		5.88		high density form		Kasper & Richards
		4.73-4.54		amorphous films		Donovan et al., A & B
Color		light grey		metallic luster		Goryunova, p. 72
Hardness		6		Mohs		Goryunova, p. 72
Knoop Microhardness	H_{25}	780 845	kg/mm^2	(110) (111)		Wolff
Cleavage		(001)		good, striated on (111)		Gmelin, p. 66
Symmetry		cubic, diamond				Donnay
Space Group		Fd3m Z-8				
Lattice Parameter		5.65754	$\overset{\circ}{A}$	high purity		Cooper
Symmetry		tetragonal		high density form obtained at P> 120 kbars		Kasper & Richards
Space Group		P422 Z-12				
Lattice Parameters	a_o	5.93	$\overset{\circ}{A}$			
	c_o	6.98				
Melting Point		941°	°C	ductile above 500°C		Gallagher
Phase Boundary		103	kbar	triple point: cubic crystal-liquid-metal	600°C	Bundy
Heat of Fusion		8.1	kcal/g.at.			Gmelin, p. 90
Boiling Point		2980	°K			
Heat of Vaporization		68	kcal/g.at.			
Heat of Sublimation		84			298	
Specific Heat		14×10^{-5}	cal/g.at. °K	high purity single crystal	2.5	Flubacher et al.
		1.067 3.302 5.033			40 100 200	Flubacher et al., Piesbergen
		5.590			300	Flubacher et al., Gerlich et al.
		5.575 5.939 6.253			310.8 499.4 762.9	Leadbetter & Settatree
		7.06			900	Gerlich et al.
		5.89 6.44		amorphous	300 500	Chen & Turnbull
Entropy		7.432	cal/g.at. °K		298.15	Flubacher et al.
Debye Temperature		374	°K		0	Flubacher et al.

PHYSICAL PROPERTIES	SYMBOL	VALUE	UNIT	NOTES	TEMP.(°K)	REFERENCES
Debye Temperature		288	°K		12	Piesbergen
		257			25	
		367			100	
		370			200	
		348			273	
		354			300	Flubacher et al.
Atomic Volume		13.5	cm^3/g.at.			Gmelin, p. 59

Thermal Conductivity

	n-type	p-type	UNIT	NOTES	TEMP.(°K)	REFERENCES
	10^{-3}	5×10^{-4}	W/cm °K	high purity single crystal,	0.2	Carruthers et al.
	2×10^{-2}	10^{-2}		$n = \sim 10^{13} - 10^{14}$	0.5	B
	1	0.5			2	
	1.8	1.2		high purity single crystal,	3	White & Woods
	11.4	9.7		$n = \sim 10^{13} - 10^{14}$	20	
	1.6	1.4			125	
		0.586	W/cm °K	high purity single crystal	298	Grieco & Montgomery, Carruthers et al. A
		5.5		initially $n = \sim 10^{12}$, after	3	Glassbrenner & Slack
		18.0		heating to melting point,	10	
		0.95		$n_p = 8 \times 10^{15}$	200	
		0.6			300	
		0.193			800	
		0.173			1210	
Thermal Expansion Coeff.		0.0062	10^{-8}/°K		2	Sparks & Swenson
		0.45			10	
		0.12			14	
		-0.40			16	
		-6.15			34	
		+2.5	10^{-6}/°K		100	Fine
		+4.0			150	
		+5.5			300	
		+6.0			550	

Elastic Coefficients

Stiffness		77°K**		273*	293**	500'			
		P=0	3×10^4 psi						
	c_{11}	13.110	13.214	12.92	12.853	12.610	10^{11} dyne/cm^2	high purity single crystal	*McSkimin **McSkimin & Andreatch A 'Beilin et al. A,B
	c_{12}	4.921	5.010	4.845	4.826	4.703			
	c_{44}	6.816	6.843	6.725	6.680	6.490			
3rd Order	c_{111}	-7.6			-7.2		10^{12} dyne/cm^2		Drabble & Fendley, McSkimin & Andreatch B
	c_{112}	-4.1			-3.8				
	c_{123}	-0.7			-0.3				
	c_{144}	0.0			-0.1				
	c_{166}	-3.1			-3.05				
	c_{450}	-0.46			-0.45				

PHYSICAL PROPERTIES	SYMBOL	VALUE	UNIT	NOTES	TEMP.($^\circ$K)	REFERENCES
Elastic Coefficients						
Compliance	s_{11}	9.787	10^{-13} cm^2/dyne			McSkimin & Andreatch A, Fine
	s_{12}	-2.672				
	s_{44}	14.971				
Stiffness						
Temperature Coeff.	Tc_{11}	-0.91	10^{-4}/$^\circ$K		80-300	McSkimin & Andreatch A
	Tc_{12}	-0.91				
	Tc_{44}	-0.92				
Shear Modulus		6.66	10^{11} dyne/cm^2		300	Fine

Young's Modulus

		73°K	300°K				
		10.45	10.25	10^{11} dyne/cm^2	(100)		McSkimin, Fine
		15.82	15.55		(111)		

PHYSICAL PROPERTIES	SYMBOL	VALUE	UNIT	NOTES	TEMP.($^\circ$K)	REFERENCES
Poisson's Ratio		0.279			73-420	Fine

Sound Velocity

		77°K	300°K	UNIT	NOTES	TEMP.($^\circ$K)	REFERENCES
Longitudinal	(001)	4.9566	4.9138	10^5 cm/sec	high purity, single crystal		McSkimin & Andreatch A
Longitudinal	(110)	5.4461	5.3996				
Shear	(001)	3.5733	3.5424				
Shear	(110)	3.5736	3.5425				
		0.027			liquid	941°C	Baidov & Gitis
		0.0265				1300°C	

PHYSICAL PROPERTIES	SYMBOL	VALUE	UNIT	NOTES	TEMP.($^\circ$K)	REFERENCES
Compressibility		1.285	10^{-12} cm^2/dyne		20°C	Fine
		2.500			941°C	Baidov & Gitis
		2.675			1300°C	

Dielectric Constant

Static ε_o

	n- (20Ω-cm)		p- (20Ω-cm)		NOTES	TEMP.($^\circ$K)	REFERENCES
	61 GHz	92 GHz	61 GHz	92 GHz			
	16.2	14.2	16.3	14.8	single crystals	300	Druesne
	15.7	13.5	15.8	13.8		77	
	16.0				pure single crystal, 9 GHz	4.2	D'Altroy & Fan B
	15.97				amorphous	300	Donovan et al. B

PHYSICAL PROPERTIES	SYMBOL	VALUE	UNIT	NOTES	TEMP.($^\circ$K)	REFERENCES
Temp. Coeff. $(1/\varepsilon_o)(d\varepsilon_o/dT)_P$		1.7	10^{-5}/$^\circ$K		77-300	Paul & Brooks
Press. Coeff. $(1/\varepsilon_o)(d\varepsilon_o/dP)_T$		-1.2	10^{-6} cm^2/kg			
Loss Tangent	tan δ	0.0003		15 GHz	4.2	Dousmanis et al.
		0.2		32 GHz	300	
		0.02		32 GHz	300	
Electrical Resistivity		10^4	ohm-cm	high purity, defect free	95	Aurich et al.
		200		n= $\sim 10^{12}$	300	

PHYSICAL PROPERTIES	SYMBOL	VALUE	UNIT	NOTES	TEMP.($^\circ$K)	REFERENCES
Electrical Resistivity		10^9		high purity, n= 10^{13}	4	Kurova &
		10^8			5	Kalashnikov A
		10^2			10	
		7			20	
		20			100	
		22			300	
		0.1		high purity	500	Morin & Maita
		0.005			800	
		0.0025			940	
Temperature Coeff.		-5.3	$10^{-2}/^\circ$K		273-323	Bullis et al.
Pressure Change		-10^6	ohm-cm	P= 120-125 kbars		Minomura & Drickamer
Electrical Resistivity		10^4		amorphous film	300	Fuhs & Stuke
		10^{10}		amorphous film, 1-6μ thick	30	Clark
		300			300	
		100		amorphous film, 0.03-0-45μ thick	300	Walley & Jonscher
Elastoresistivity	$1/X \cdot d\rho/\rho$	-6		amorphous	300	Fuhs & Stuke
Mobility						
Electron	u_n	1.2×10^6	cm^2/V sec	high purity, single crystal, $n_n = \sim 10^{12}$	1.8-4.2	Rollin & Rowell
		3×10^4		$n_n = 2 \times 10^{13}$	77	Pödör,
		4×10^3			300	Baranskii et al.
		4×10^4		$n_p = 9 \times 10^{13}$	50	Paige
		3×10^3			300	
Hole	u_p	5×10^5		$n_p = \sim 10^{13}$	10	Brown & Bray,
		4×10^4			77	Baranskii et al.
		2×10^3			300	
		8×10^4		$n_n = 7 \times 10^{12}$	50	Paige
		2×10^4			100	
		2×10^3			300	
Electron		800		5μ thick, epitaxial film	300	Marucci
		600		1μ		
		100		0.2μ		
		10^{-2}		1-20μ thick, amorphous films	300	Clark
Temperature Coeff.						
Electron		T^{-4}		$n_n = \sim 4 \times 10^{12}$	4-20	Rollin & Rowell
		$T^{-1.66}$		high purity	100-300	Morin
		$T^{-2.5}$		high purity	>400	Golikova & Petrov
Hole		$T^{-1.5}$		$n_p = 10^{13}$	<70	Brown & Bray
		$T^{-2.33}$			100-300	Brown & Bray, Morin

ELECTRICAL PROPERTIES	SYMBOL	VALUE	UNIT	NOTES	TEMP.(°K)	REFERENCES
Lifetime						
Electron	τ_n	4×10^{-2}	μ sec.	n-type single crystal	4	Ascarelli & Brown
		20			125	Iglitsyn & Yurova
		200			300	
		30-120		p-type	90	Buschor &
		2×10^4		p-type	300	Baldinger, Sloan & Hauser
Hole	τ_p	0.03		p-type	90	Buschor &
		4×10^4		n-type	300	Baldinger, Sloan & Hauser
Cross-Section						
Electron	σ_n	10^{-12}	cm^2	high purity n-type, single crystal	4	Ascarelli & Brown, Koenig et al.
	σ_n	4×10^{-16}		lithium drift diode	90	Buschor & Baldinger
Hole	σ_p	3×10^{-14}				
Piezoelectric Coeff.	d	6	10^{-11}C/N	n_n, $n_p = \sim10^{13}$	300	Gundjian
Electrostriction		1.6-7.2	$10^{-11}cm^2/V^2$		300	Gundjian

Piezoresistance

		n- (16.6Ω-cm)	p- (15Ω-cm)	UNIT	NOTES	TEMP.(°K)	REFERENCES
	π_{11}	-5.2	-10.6	$10^{-12}cm^2/dyne$	n-, As-doped, p-, Ga-doped, $P = 10^7 - 10^8$ dyne/cm^2	300	Smith
	π_{12}	-5.5	+ 5.0				
	π_{44}	-38.7	+98.6				

	77°K		300°K		UNIT	NOTES		REFERENCES
P=	2×10^8	4×10^9	2×10^8	2.5×10^9	dyne/cm^2			
	170	500	10	140	$10^{-12}cm^2/dyne$	$n_p = 5\times10^{13}$		Asche et al.
	90	90	50	50		$n_p = 2\times10^{17}$		

Elastoresistance Coefficient

		n-type	p-type			TEMP.(°K)	REFERENCES
	m_{11}	20.5	-4.3			300	Smith
	m_{12}	20.3	+8.3				
	m_{44}	-93.4	+66.5				

Effective Mass

	SYMBOL	m_t	m_ℓ	UNIT	NOTES	TEMP.(°K)	REFERENCES
Transverse Electron	m_t	0.08152	1.588	m_o	cyclotron resonance in high purity single crystal, 24 GHz.	1	Levinger & Frankl
Longitud. Electron	m_ℓ						
		0.079	1.74		magnetoabsorption	1.7	Halpern & Lax
		0.0819	1.64		cyclotron resonance	4	Dexter et al.
		0.079	1.54		magnetopiezotransmission in high purity single crystal, 90 kGauss, (L_1 Conduction Band)	20	Aggarwal et al.

ELECTRICAL PROPERTIES	SYMBOL	VALUE (100)	(111)	(110)	UNIT	NOTES	TEMP.(°K)	REFERENCES
Effective Mass								
Heavy Electron	m_1	0.131	0.197	0.340		magnetoabsorption at 74 kGauss on high purity single crystals	1.7	Halpern & Lax
Light Electron	m_2		0.079	0.101				
	m_1			0.352		magnetopiezotransmission	20	Aggarwal et al.
	m_2			0.0951				
	m_1	0.1346	0.2057	0.3598		cyclotron resonance at 24 GHz on high purity crystals	1	Levinger & Frankl
	m_2		0.08152	0.09875				
Electron (at direct transition)	m_n	0.041			m_o	electromagnetoreflectivity at 0.13μ, (111)-oriented	1.5	Rouzeyre et al.
		0.042				electromagnetoreflectivity	300	Groves et al.
Electron (at indirect trans.)		0.128				high purity, single crystal magnetopiezotransmission	0	Aggarwal et al. B
		0.1346				cyclotron resonance meas.	1.0	Levinger & Frankl
		0.135				Faraday rotation meas.	300	Ukhanov & Maltsev
Hole	m_p	0.34				electrical meas.	20-300	D'Altroy & Fan A
Hole		(100)	(111)	(110)				
Light Hole	m_{lp}		0.043			electromagnetoreflectivity	1.5	Rouzeyre et al.
Heavy Hole	m_{hp}		0.364					
	m_{lp}	0.04368	0.04187	0.04236		cyclotron resonance in high purity single crystals at 24 GHz.	1	Levinger & Frankl
	m_{hp}	0.2825	0.377					
	m_{hp}			0.352			4	Dexter et al.
Split-off Valence Band	m_{so}	0.084				magnetoelectroreflectivity	300	Groves et al.

Diffusion Coeff. and Energy Levels	Dopant	D_o	D	E_{act}	E_d	E_a		Temp.(°C)	
		(cm²/sec)			(eV)				
	Ag	4.4×10^{-2}		1.0				70-940	Bugai et al. A
			8.8×10^{-7}					800	
					0.14 VB	elec. meas.		300	Tyler A
					0.25 CB				
					0.10 CB				
	Al		1.5×10^{-14}					750	Meer & Pommerenig
			3×10^{-13}					850	
					0.0102	elec. meas.		50-300°K	Geballe & Morin
					0.0108	optical meas.		9°K	Jones & Fischer A
	As	1.503		2.39				580-650	Wölfle & Dorendorf
			8×10^{-15}					580	
			4.7×10^{-14}					650	
					0.0127	elec. meas.		300°K	Geballe & Morin

ELECTRICAL PROPERTIES Diffusion Coeff. and Energy Levels	Dopant	D_0 (cm²/sec)	D (cm²/sec)	E_{act} (eV)	E_d (eV)	E_a (eV)	NOTES	TEMP. (°C)	REFERENCES
	Au	2.25×10^2		2.5				600-900°K	Dunlap B
			1.5×10^{-12}					600°K	
			4×10^{-9}					900°K	
			10^{-12}				irradiated crystals	300°K	Klimkova & Nuyazova
					0.041 VB	0.15 VB	p-type, electrical meas.	4-300°K	Ostroborodova
						0.15 VB	n-type	4-300°K	Ostroborodova,
						0.21 CB		77-400°K	Dunlap A
						0.41 CB			
	B	1.8×10^9		4.55				600-900°K	Dunlap A
			7.4×10^{-13}					800°K	
			5×10^{-19}					900°K	Meer & Pommerenig
			10^{-15}					750	
						0.0104	elec. meas.	50-300°K	Geballe & Morin
						0.01047	photocond. meas.	7°K	Sidorov & Lifshits
						0.0105	opt. absorption	10°K	Jones & Fisher A
	Be	0.5		2.5				720-920	Belyaev & Zhidkov
			2×10^{-13}					700	
			2×10^{-11}					900	
					0.067		photocond. meas.	6°K	Tyapkina
					0.026		opt. & elec.	300°K	Shenker et al.
	Bi	3.3		2.47				650-850	Sharma, p. 89
			8.1×10^{-12}					800	
					0.012			4-20°K	Kurova & Kalashnikov B
	Cd	1.7×10^9		4.4				760-910	Kosenko
			10^{-12}					760	
			5×10^{-10}					910	
						0.045VB	elec. meas.	78-300°K	Iglitsyn & Yurova
						0.16 VB			
	Co	4.4×10^{-3}		0.88			high-defect crystal	750-850	Wei
		0.16		1.12			low-defect crystal		
			8.7×10^{-7}					800	
						0.25	elec. meas.	100-300°K	Tyler et al. B
						0.31			
	Cr					0.12 VB			Tyler B

ELECTRICAL PROPERTIES	Dopant	D_o	D	E_{act}	E_d	E_a	NOTES	TEMP.(°C)	REFERENCES
		(cm^2/sec)		(eV)					
Diffusion Coeff. and Energy Levels	Cu	4×10^{-3}		0.33			interstitial diffusion in p-type crystals	350-750 / 350 / 725	Hall & Racette
			10^{-5}						
			10^{-4}						
			$10^{-9}\text{-}10^{-10}$				substitutional diffusion in n-type crystals of increasing carrier concentration	650	
						0.04 VB	elec. meas.	4-300°K	Woodbury & Tyler B, Ivanov
						0.32 VB			
						0.26 CB			
Temperature Coeff.	$dE_a/dT = -5.3 \times 10^{-4}$ eV/°K						photocond. meas.	4-150°K	Ivanov
	Fe	0.13		1.08				750-940	Bugai et al. B
			2.3×10^{-7}					800	Wei
						0.34 VB	elec. meas.	77-300°K	Tyler & Woodbury
						0.27 CB			
	Ga	34		3.1				600-900	Dunlap A
			8.9×10^{-14}					800	
						0.0109	elec., opt., photocond. meas.	6, 9°K / 50-100°K	Jones & Fischer A, Kurova et al., Geballe & Morin
	Ge	7.8		2.98				766-928	Letaw et al.
			3.5×10^{-14}					766	
			2×10^{-12}					928	
	H		5×10^{-5}	1.01				780-930	Van Wieringen & Warmoltz
	He	6.1×10^{-3}		0.69				795-872	Van Wieringen & Warmoltz
	Hg					0.16 VB	elec. meas.	77°K	Woodbury & Tyler A
						0.37			
						0.091 VB	optical meas.	10°K	Chapman & Hutchinson A,B
						0.088		35°K	Aranovich et al., Kurova et al.
						0.087	elec. & photocond. meas.	4-30	Borello & Levinstein
						0.23 VB			
	In	10		2.77			vol. diffusion	510-880	Panteleev
		6.8×10^6		2.8			dislocation diffusion		
			9.5×10^{-13}					800	
		5.8×10^3		2.47			electron-irrad.	650-850	Dudko et al.
						0.0116	optical meas.	6°K	Jones & Fischer A
							photocond. meas.	8°K	Sidorov & Lifshits
						0.0112	elec. meas.	50-100°K	Geballe & Morin

ELECTRICAL PROPERTIES — Diffusion Coeff. and Energy Levels

Dopant	D_o (cm^2/sec)	D (cm^2/sec)	E_{act} (eV)	E_d (eV)	E_a (eV)	NOTES	TEMP.(°C)	REFERENCES
Li	9.1×10^{-3}		0.57				300-400	Pratt & Friedman
		5×10^{-7}					400	
				0.0099		lumines. meas.	4	Aggarwal et al. A
Ni	0.8		0.9				700-850	Van der Maesen & Brenkman
		1.5×10^{-5}					700	
		5.5×10^{-5}					875	
					0.23 VB 0.31 CB	elec. and photocond. meas.	77-350	Tyler et al. A, Gouskov et al.
O	0.4		2.076				800-900	Corbett et al.
		6.7×10^{-11}					800	
				0.1		optical meas.	90, 300°K	
P	3.3		2.5				600-900	Dunlap A
		10^{-11}					800	Sharma, p. 89
				0.012		elec. meas.	50-200°K	Geballe & Morin
Pb		2×10^{-14}	3.6				800	Boltaks, p. 164, 167
Pt					0.040	elec. meas.	15-400°K	Dunlap C
					0.2 CB	photocond. meas.	300°K	Slatter, Dunlap C
S		10^{-9}		0.18		elec. meas.	920	Tyler B
Sb	1.3		2.26				800-900	Miller & Smits, Bösenberg
		3.5×10^{-11}					800	
		2.3×10^{-10}					900	
	21		2.1			electron irrad.	650-850	Dudko et al.
				0.0097		elec. meas.	50-200°K	Geballe & Morin
Se		10^{-10}					920	Tyler B
				0.14 CB 0.28 CB		elec. meas.		Tyler B
Sn	1.7×10^{-2}		1.9					Boltaks, p. 167
		2×10^{-11}					800	
					0.508	lumines. meas. (exciton-tin center)	77°K	Hergenrother & Feldman
					0.551	optical meas.	300	Pankove
Te	5.6		2.43			bulk diffusion	750-900	Ignatkov & Kosenko
		3×10^{-11}					800	
	2.0		2.82			surface diff.	750-900	
		10^{-13}					800	
		10^{-11}				surface diff.	920	Tyler B

ELECTRICAL PROPERTIES	Dopant	D_o	D	E_{act}	E_d	E_a	NOTES	TEMP.(°C)	REFERENCES
		(cm^2/sec)		(eV)					
Diffusion Coeff. and Energy Levels	Th	1.7×10^3		3.4				800-930	Tagirov & Kuliev
			1.5×10^{-13}					800	
			9×10^{-12}					930	
						0.0131	optical meas.	10°K	Jones & Fischer A
	Zn	5		2.7				600-900	Dunlap A
			10^{-11}					700	
			5×10^{-14}					900	
						0.029 VB	elec. and	4-400°K	Dunlap C,
						0.0827 VB	optical meas.		Sidorov et al.,
									Bray & Pinson,
									Fisher & Fan

ELECTRICAL PROPERTIES	SYMBOL	VALUE	UNIT	NOTES	TEMP.(°K)	REFERENCES
Energy Gap						
Direct ($\Gamma_{25'}$-$\Gamma_{2'}$)	E_o	0.889	eV	electromagnetoreflectivity at 0.13 to 0.14µ, 50 kG	1.5	Rouzeyre et al.
		0.898		magnetoabsorption in high	1.5-4.2	Zwerdling et al.
		0.889		purity single crystal	77	
		0.805		electroreflectivity	300	Lukes & Schmidt, Seraphin et al.
				Faraday rotation		Piller & Patton
				electromagnetoreflectivity		Groves et al.
				optical transmission		Hobden
		0.8807		electric field modulation	89	Frova & Penchina
		0.7990			305	
Spin-orbit Splitting at $\Gamma_{25'}$ (Γ_7^+-Γ_8^+)	Δ_o	0.282		electromagnetoreflectivity	300	Groves et al.
		E_o Δ_o				
		0.882 0.292		electroreflectivity at 0.48 to 1.7µ	24	Nishino & Hamakawa
		0.872 0.292			83	
		0.836 0.297			209	
		0.796 0.296			300	
	E_o	0.840		optical absorption in heavily doped single crystals, $n_n = 10^{19}$	80	Haas
		0.764			295	
Indirect ($\Gamma_{25'}$-L_1)	E_g	0.744		magnetoabsorption	1.5	Zwerdling et al.
		0.77		magnetopiezotransmission on high purity crystals	20	Aggarwal et al.
		0.6643		piezotransmission	298	Engeler et al.
		0.7410		optical absorption in high purity crystals	4	Macfarlane et al. A
		0.7403			20	
		0.7339			77	
		0.6640			291	
		0.6153			416	
		0.673		optical absorption in heavily doped crystals	80	Haas
		0.624			295	

ELECTRICAL PROPERTIES	SYMBOL	VALUE		UNIT	NOTES	TEMP.($^\circ$K)	REFERENCES
Energy Gap							
L_1-Δ_1		0.18		eV		300	Jayaraman
$L_{3'}$-L_1		2.05			electroreflectivity	300	Ghosh
Spin-orbit Splitting at $L_{3'}$		0.19			at low field		
		0.17			anisotropic in polarized light		
Λ_3-Λ_1	E_1						
Spin-orbit Splitting at Λ_3	ΔE_1	E_1	ΔE_1				
		2.22	0.19	eV	optical absorption	120	Potter, Ghosh,
		2.13	0.20		electroreflectivity	300	Cardona et al. A
		2.223	0.189		electroreflectivity	24	Nishino &
		2.212	0.189			83	Hamakawa
		2.155	0.187			209	
		2.09	0.170			300	
$\Gamma_{25'}$-Γ_{15}	$E_{0'}$						
$\Gamma_{25'}(\Gamma_8+)$-$\Gamma_{15}(\Gamma_6-)$		2.8			electroreflectivity	300	Ghosh
$\Gamma_{25'}(\Gamma_8+)$-$\Gamma_{15}(\Gamma_8-)$		2.93					
$\Gamma_{25'}(\Gamma_7+)$-$\Gamma_{15}(\Gamma_6-)$		3.09					
$\Gamma_{25'}(\Gamma_7+)$-$\Gamma_{15}(\Gamma_8-)$		3.22					
Spin-orbit Splitting at Γ_{15}		0.13					
Spin-orbit Splitting at $\Gamma_{25'}$		0.29					
Δ_5-Δ_1		3.13					
X_4-X_1	E_2	4.49			electroreflectivity	5	Yu & Cardona
or							
$(\Sigma_2$-$\Sigma_3)$		4.42			electroreflectivity	300	Cardona et al. A
		4.39			electroreflectivity	300	Ghosh
$L_{3'}$-L_3	$E_{1'}$	5.35			electroreflectivity	300	Ghosh
Δ_5-$\Delta_{2'}$		7.2			optical reflectivity in high purity crystals	300	Pajasova
Energy Gap Temperature Coeff.							
Direct	dE_0/dT	-4.4		10^{-4}eV/$^\circ$K	optical absorption in high purity, strain and dislocation free single crystal	20-300	Harbeke
		-3.9			magnetoabsorption in high purity, single crystal	77-293	Zwerdling et al.
		-1.2				4-77	
Indirect	dE_g/dT	-3.7			optical absorption	4-291	Macfarlane et al. B
	dE_2/dT	-2.4			electroreflectivity	5	Yu & Cardona
		-1.8			optical meas.	85-300	Cardona & Sommers
	$d(L_{3'}$-$L_1)/dT$	-4.2			reflectivity meas.	85-300	Cardona & Sommers

ELECTRICAL PROPERTIES	SYMBOL	VALUE	UNIT	NOTES	TEMP.($^\circ$K)	REFERENCES
Energy Gap						
Pressure Coeff.						
Direct	dE_o/dP	12.5	10^{-6}eV/kg cm^{-2}	piezoelectroreflectance (111) and (110)	300	Pollack & Cardona
Indirect	$d(\Gamma_{25'}-L_1)/dP$	5		electrical meas. at 35 kbars	300	Jayaraman et al., Arizumi et al.
	$d(\Gamma_{25'}-\Delta_1)/dP$	-1.3				
		-1.87		electrical meas.	4.2	Arizumi et al.
	dE_1/dP	7.5		optical reflectivity	300	Zallen & Paul
	dE_2/dP	5.6				
Volume Coefficient		3.7	eV	electrical meas.	300	Jayaraman et al.
Energy Gap		0.6	eV	optical meas. on amorphous films		Donovan et al. B
Deformation Potential						
Conduction Band						
Shear	Ξ_u	19.3	eV	cyclotron resonance meas. on ultra high purity crystals	4	Murase et al.
Dilatational	Ξ_d	-12.3				
	Ξ_u	16.3		optical stress modulation	80	Balslev A
		15.9			297	
	Ξ_u	18	eV	piezobirefringence	77	Riskaer
				optical meas.	300	Remenyuk et al.
	Ξ_u	16.3		drift mobility meas.	77-300	Schetzina & McKelvey
Valence Band	a	2.6		electrical meas.	4.2	Arizumi et al.
	b	-2.6(001)		piezoelectroreflectance	300	Pollack & Cardona,
	d	-4.7(111)			77	Gavini & Cardona
	b	-2.4		piezoreflectance, for (001) and (111)	100,300	Balslev C,
	d	-4.1			4	Jones & Fischer B
Work Function	ϕ	4.80	eV	photoelectric meas. on cleaved (111) surface	300	Gobeli & Allen
	Na K Cs 3.6 2.6 2.2		eV	0.3% coverage		Weber & Peria, Palmberg
Photoelectric Threshold	Φ	4.80	eV	photoelectric meas.	300	Gobeli & Allen
Electron Affinity	ψ	4.13	eV	photoelectric meas.	300	Gobeli & Allen
Barrier Heights		0.37	eV	Cu, Ru		Jaeger & Kosak
		0.40		Rh, Os		
		0.42		Fe, Ir		
		0.45		Pd, Pt		
		0.47		Ag, Au		
		0.51		Co		
		0.57		Ni		

GERMANIUM

ELECTRICAL PROPERTIES	SYMBOL	VALUE	UNIT	NOTES	TEMP.($^\circ$K)	REFERENCES
Electron Secondary Emission Coeff.		1.08		pure single crystal	300	Bolshov & Seleznev
		0.98		liquid	1232	
				maximum value at 550 V		
Richardson's Constant		0.023	Amp/cm^2 $^\circ$K^2		1232	Bolshov
Phonon Branch Spectra						
Raman Phonon		37.3	meV	piezotransmission Raman scattering	300	Engeler et al., Parker et al.
Transverse Acoustic	TA	7.76		optical absorption in high purity single crystals; electroabsorption at 77°K, and piezotransmission at 300°K	77-291	Macfarlane et al. A, Zemskii & Mochan, Engeler et al., Frova et al.
Longitudinal Acoustic	LA	27.57				
Longitudinal Optic	LO	30.16			195	
		31.02			363	
Transverse Optic	TO	36.19			195	
		36.71			363	
Pressure Coeff.	dTO/dP	5.6	10^{-6}meV/kg cm^{-2}	elec. meas. at 18 kbar	1.3	Zavaritskii
	dLO/dP	6.2				
	dTA/dP	-1.64				
	dLA/dP	1.49				
Seebeck Coefficient		-0.9	mV/$^\circ$K	high purity n-type, crystal	293	Domenicali, Glazov et al.
		-0.2			873	
		0.1		liquid	1223	

	n-type	p-type	UNIT	NOTES	TEMP.($^\circ$K)	REFERENCES
	-5.0	+5.0	mV/$^\circ$K	high resistivity crystals	40	Freud & Rothberg
	-1.2	+1		$n_n = 10^{14}$, As-doped	200	Geballe & Hull
	-1.1	+1		$n_p = 10^{14}$, Ga-doped	300	
	-0.8	0			330	
	-0.5	-0.4			375	
	0.24	0.23		$n = 10^{19}$	700	Beers et al.
	0.15 to 0.40			p-type film	300	Hui & Corra
	0.2			amorphous	300	Grigorivici
Nernst-Ettingshausen Coeff.	-0.55	+0.01	cgs	single crystals,	150	Bashirov & Tsidilkovskii
	-0.1	+0.05		n-type resistivity= 3 Ω-cm	300	
		0		p-type resistivity= 0.18	350	
	-1.15				400	
		-0.55			500	
	-0.12	-0.4			600	

	VALUE	UNIT	NOTES	TEMP.($^\circ$K)	REFERENCES
Nernst Coefficient	10.5	10^{-8}V/$^\circ$K G^{-1}	$n_n = 10^{13}$, H= 3 kGauss	77	Herring et al.
Magnetic Susceptibility	0.105	10^{-6} cgs		290	Busch & Yuan
	0.116		solid	940°C	Glazov & Chizhevskaya
	0.088		liquid	950°C	
g-Factor					
Electron g_n	1.8		magnetoabsorp. at 74 kG	1.7	Halpern & Lax
Hole g_p	7.2		cyclotron resonance	4	Fujiyasu et al., Hensel
Superconducting Transition Temp.	5.4	$^\circ$K	P= 120 kbar		Buckel & Wittig

GERMANIUM

OPTICAL PROPERTIES	SYMBOL	VALUE	UNIT	WAVELENGTH (μ)	NOTES	TEMP.(°K)	REFERENCES
Absorption Coefficient		10^4	cm^{-1}	1.5	high quality film	300	Wales et al.
		10^2		5			
Transmission		60	%	0.04	700 Å thick film	50	Hunter et al.
		40		1.5-15	single crystal	300	Linsteadt
Refractive Index		0.8		0.05	films	300	Marton & Toots, LaVilla & Mendlowitz
		0.5		0.16			
		4.3		2.0	high quality film	300	Wales et al.
		4.2		0.4	single crystals measured in N atmos.	300	Archer
		4.5		0.5			
		5.6		0.6			
		4.1254		2.0		300	Rank et al.
		4.1016		2.06	single crystals	300	Salzberg & Villa
		4.0170		4.87			
		4.0018		12.20			
		4.0012		16.00			
		3.98		80-140	high purity crystal	7-300	Aronson
		3.99		3-12	amorphous	300	Donovan et al. B
		4.73		0.78			
Temperature Coeff.	$(1/n)(dn/dT)$	1.8	10^{-4}/°K	1	films	300-520	Lukes
		1.432		1.934	single crystal	300	Rank et al., Lukes
		1.287		2.174			
		1.280		2.246			
		1.231		2.401			
		0.69		5		87-300	Cardona et al. B
Pressure Coeff.	$(1/n)(dn/dP)$	-0.7	$10^{-6}cm^2$/kg		single crystal	300	Paul & Brooks
Spectral Emissivity		0.65		0.65		700-1150	Brekhovskikh
		0.58			at melting point	1200	
		0.18			liquid		Allen
Birefringence	$\varepsilon_\parallel - \varepsilon_\perp / X$	0.2	$10^{-11}cm^2$/dyne	2.07	stress on (100)	300	Higginbotham et al.
		-1.2		1.70			
		2.8		2.07	stress on (111)		
		2.2		1.65			
Stress Optical Constant	Q_1	-0.61	°K cm/kg	2.2	single crystal	300	Schmidt-Tiedemann
		+5.7					
Photoelastic Tensor Component	p_{11}	0.27		10.6	high purity, single crystal	300	Abrams & Pinnow
	p_{12}	0.235					
	p_{44}	0.125					
Dispersion	$dn/d\lambda$	0.035	λ^{-1}	2.9		300	Billard & Hily
		0.012		4.0			
Lases		5178	Å		100 milliAmps	900-1000°C	Silfvast et al.

ABRAMS, R.L. and D.A. PINNOW. Acousto-Optic Properties of Crystalline Germanium. J. OF APPLIED PHYS., v. 41, no. 7, June 1970. p. 2765-2768.

AGGARWAL, R.L. et al. Excitation Spectra of Lithium Donors in Silicon and Germanium. PHYS. REV., v. 138, no. 3A, May 1965. p. A882-A893.   [A]

AGGARWAL, R.L. et al. Magnetopiezotransmission Studies of the Indirect Transition in Germanium. PHYS. REV., v. 180, no. 3, Apr. 1969. p. 800-813. [B]

ALLEN, F.G. Emissivity at 0.65 Micron of Silicon and Germanium at High Temperatures. J. OF APPLIED PHYS., v. 28, no. 12, Dec. 1957. p. 1510-1511.

ARANOVICH, R.M. et al. Photoelectric Properties of Mercury-Doped Germanium. SOVIET PHYS. SEMICONDUCTORS, v. 1, no. 4, Oct. 1967. p. 502-504.

ARCHER, R.J. Optical Constants of Germanium: 3600-7000 Å. PHYS. REV., v. 110, no. 2, Apr. 1958. p. 354-358.

ARIZUMI, T. et al. Uniaxial Stress Effect on (000) and (100) Conduction Band Minima of Germanium. JAPAN. J. OF APPL. PHYS., v. 8, no. 6, June 1969. p. 700-703.

ARONSON, J.R. et al. Low-Temperature Far-Infrared Spectra of Germanium and Silicon. PHYS. REV., v. 135, no. 3A Aug. 1964. p. A785-A788.

ASCARELLI, G. and S.C. BROWN. Recombination of Electrons and Sonors in n-Type Germanium. PHYS. REV., v. 120, no. 5, Dec. 1960. p. 1615-1626.

ASCHE, M. et al. Piezoresistivity of p-Type Germanium (In Ger.). PHYS. STATUS SOLIDI, v. 18, no. 2, 1966. p. 749-754.

AURICH, R. et al. High-Purity Germanium-Surface Conductivity and Carrier Lifetime. ANN. DER PHYSIK, Leipzig, ser. 7, v. 11, July 1963. p. 83-100.

BAIDOV, V.V. and M.B. GITIS. Velocity of Sound in and Compressibility of Molten Germanium and Silicon. SOVIET PHYS. SEMICONDUCTORS, v. 4, no. 5, Nov. 1970. p. 825-826.

BALSLEV, I. Indirect Absorption in Germanium Under Combined Static and Oscillatory Stress. PHYS. LETTERS, v. 24A, no. 2, Jan. 1967. p. 113-114. [A]

BALSLEV, I. Inter-Valence-Band Transitions in Uniaxially Stressed Germanium and Gallium Arsenide. PHYS. REV., v. 177, no. 3, Jan. 1969. p. 1173-1178. [B]

BALSLEV, I. Direct Edge Piezoreflectance in Germanium and Gallium Arsenide. SOLID STATE COMM. v. 5, no. 4, Apr. 1967. p. 315-317. [C]

BARANSKII, P.I. et al. Study of Magnetoelectric Properties in Pure Elemental Monocrystalline Germanium (In Russ.). AKAD. NAUK SSSR., IZV. NEORGAN. MAT., v. 3, no. 10, Oct. 1967. p. 1782-1786.

BASHIROV, R.I. and I.M. TSIDILKOVSKII. The Nernst-Ettingshausen Effect in Germanium. SOVIET PHYS. TECH. PHYS., v. 1, no. 10. 1957. p. 2129-2133.

BEERS, D.S. et al. Thermal Conductivity of Germanium, Silicon and III-V Compounds at High Temperatures. INTERNAT. CONF. ON THE PHYS. OF SEMICONDUCTORS, PROC., Exeter, July 1962. Ed. by STICKLAND, A.C. London, Inst. of Phys. and the Phys. Soc., 1962. p. 41-48.

BEILIN, V.M. et al. Elastic Constants of Heavily Doped n-Type Silicon and p-Type Germanium. SOVIET PHYS. SOLID STATE, v. 12, no. 3, Sept. 1970. p. 531-535. [B]

BEILIN, V.M. et al. Elastic Moduli of Heavily Doped n-Type Germanium at 300-550°K. SOVIET PHYS. SOLID STATE, v. 10, no. 5, Nov. 1968. p. 1226-1227. [A]

BELYAEV, Yu. I. and V.A. ZHIDKOV. The Diffusion of Beryllium in Germanium. SOVIET PHYS. SOLID STATE, v. 3, no. 1, July 1961. p. 133-134.

BILLARD, P. and C. HILY. Weak Field Optical Systems for Use in the Infrared (In Fr.). ACTA ELECTRONICA, v. 11, no. 3, July 1968. p. 237-325.

BOLSHOV, V.G. Thermionic Emission at the Solid-Liquid Transition Point. SOVIET PHYS. TECH. PHYS., v. 1, no. 6, June 1957. p. 1123-1133.

BOLSHOV, V.G. and V.K. SELEZNEV. Secondary Electron Emission in Copper, Germanium and Tin in the Solid and Liquid States. SOVIET PHYS. TECH. PHYS., v. 1, no. 8, 1957. p. 1612-1618.

BOLTAKS, B.I. Diffusion in Semiconductors. N.Y. Academic Press, 1963. 378 pp.

BORRELLO, S.R. and H. LEVINSTEIN. Preparation and Properties of Mercury-Doped Germanium. J. OF APPLIED PHYS.,
v. 33, no. 10, Oct. 1962. p. 2947-2950.

BÜSENBERG, W. Diffusion of Antimony, Arsenic and Indium in Solid Germanium (In Ger.). ZEITSCHRIFT FUER NATUR-
FORSCHUNG, v. 10a, no. 4, Apr. 1955. p. 285-291.

BRATTAIN, W.H. and H.B. BRIGGS. The Optical Constants of Germanium in the Infrared and Visible. PHYS. REV.,
v. 75, no. 11, June 1949. p. 1705-1710.

BRAY, R. and W.E. PINSON. Determination of the Hot Carrier Distribution Function from Anisotropic Infrared
Absorption in p-Type Germanium. PHYS. REV. LETTERS, v. 11, no. 6, Sept. 1963. p. 268-271.

BREKHOVSKIKH, V.F. Experimental Determination of the Emissive Power of Germanium and Silicon in the Temperature
Range 700-1200°K. Progress in Heat Transfer. Ed. by P.K. KONAKOV. N.Y. Consultants Bureau, 1966. p. 145-150.

BROWN, D.M. and R. BRAY. Analysis of Lattice and Ionized Impurity Scattering in p-Type Germanium. PHYS. REV.,
v. 127, no. 5, Sept. 1962. p. 1593-1602.

BUCKEL, W. and J. WITTIG. Superconductivity of Germanium and Silicon under High Pressure (In Ger.). PHYS.
LETTERS, v. 17, no. 3, July 1965. p. 187-188.

BUGAI, A.A. et al. Diffusion and Solubility of Silver in Germanium. SOVIET PHYS. TECH. PHYS. v. 2, no. 8,
Aug. 1957. p. 1553-1557. [A]

BUGAI, A.A. et al. Diffusion and Solubility of Iron in Germanium. SOVIET PHYS. TECH. PHYS., v. 2, no. 1,
Jan. 1957. p. 183-184. [B]

BULLIS, W.M. et al. Temperature Coefficient of Resistivity of Silicon and Germanium near Room Temperature.
SOLID STATE ELECTRONICS, v. 11, no. 7, July 1968. p. 639-646.

BUNDY, F.P. Phase Diagrams of Silicon and Germanium to 200 kbar, 1000°C. J. OF CHEM. PHYS., v. 41, no. 12,
Dec. 1964. p. 3809-3814.

BUSCH, G. and S. YUAN. Magnetic Susceptibility of Molten B-Elements. PHYS. KONDENS. MATERIE, v. 1, no. 1,
1963. p. 37-66.

BUSCHOR, F. and E. BALDINGER. Study of Carrier Recombination in Intrinsic Germanium at Low Temperatures
(In Ger.). HELV. PHYS. ACTA, v. 43, no. 2, Jan. 1970. p. 133-161.

CARDONA, M. and H.S. SOMMERS. Effect of Temperature and Doping on the Reflectivity of Germanium in the Fundamental Absorption Region. PHYS. REV., v. 122, no. 5, June 1961. p. 1382-1388.

CARDONA, M. et al. Electroreflectance at a Semiconductor-Electrolyte Interface. PHYS. REV., v. 154, no. 3,
Feb. 1967. p. 696-720. [A]

CARDONA, M. et al. Dielectric Constant of Germanium and Silicon as a Function of Volume. PHYS. AND CHEM. OF
SOLIDS, v. 8, Jan. 1959. p. 204-206. [B]

CARRUTHERS, J.A. et al. The Thermal Conductivity of Germanium and Silicon between 2 and 300°K. ROYAL SOC. OF
LONDON, PROC., A, v. 238, no. 1215, Jan. 1957. p. 502-514. [A]

CARRUTHERS, J.A. et al. Thermal Conductivity of p-Type Germanium between 0.2 and 4°K. CRYOGENICS, v. 2, no. 3,
Mar. 1962. p. 160-166. [B]

CHAPMAN, R.A. and W.G. HUTCHINSON. Excitation Spectra and Photo-Ionization of Neutral Mercury Centers in
Germanium. PHYS. REV., v. 157, no. 3, May 1967. p. 615-622.   [A]

CHAPMAN, R.A. and W.G. HUTCHINSON. Excited States of Mercury-Induced Double Acceptors in Germanium. SOLID STATE
COMM., v. 3, no. 9, Sept. 1965. p.293-296.   [B]

CHEN, H.S. and D. TURNBULL. Specific Heat and Heat of Crystallization of Amorphous Germanium. J. OF APPLIED PHYS.,
v. 40, no. 10, Sept. 1969. p. 4214-4215.

CHOPRA, K.L. and S.K. BAHL. Structural, Electrical and Optical Properties of Amorphous Germanium Films. PHYS. REV.,
B, Ser. 3, v. 1, no. 6, Mar. 1970. p. 2545-2556.

CLARK, A.H. Electrical and Optical Properties of Amorphous Germanium. PHYS. REV., v. 154, no. 3, Feb. 1967.
p. 750-757.

COOPER, A.S. Precise Lattice Constants of Germanium, Aluminum, Gallium Arsenide, Uranium, Sulphur, Quartz and
Sapphire. ACTA CRYSTALLOGRAPHICA, v. 15, 1962. p. 578-582.

CORBETT, J.W. et al. The Configuration and Diffusion of Isolated Oxygen in Silicon and Germanium. J. PHYS. AND
CHEM. OF SOLIDS, v. 25, 1964. p. 873-879.

D'ALTROY, F. and H.Y. FAN. Effective Mass of Carriers and Relaxation Time in Germanium. AMER. PHYS. SOC., BULL., ser. 1, v. 30, no. 2, Mar. 1955. p. 38   [A]

D'ALTROY, F. and H.Y. FAN. Effect of Neutral Impurity on the Microwave Conductivity and Dielectric Constant of Germanium at Low Temperatures. PHYS. REV., v. 103, no. 6, Sept. 1956. p. 1671-1674.   [B]

DEXTER, R.N. et al. Cyclotron Resonance Experiments in Silicon and Germanium. PHYS. REV., v. 104, no. 3, Nov. 1956. p. 637-644.

DOMENICALI, C.A. Experimental Study of Thermoelectric Power and Resistivity of Solid and Liquid Germanium in the Vicinity of Its Melting Point. J. OF APPLIED PHYS., v. 28, no. 7, July 1957. p. 749-753.

DONNAY, J.D.H. (Ed.) Crystal Data. Determinative Tables. 2nd Ed. American Crystallographic Assoc. 1963.

DONOVAN, T.M. et al. A High Density Form of Amorphous Germanium. PHYS. LETTERS, v. 32A, no. 2, June 1970. p. 85-86. [A]

DONOVAN, T.M. et al. Optical Properties of Amorphous Germanium Films. PHYS. REV., B, ser. 3, v. 2, no. 2, July 1970. p. 397-413. [B]

RCA. DAVID SARNOFF RES. CENTER. Negative Mass Carrier Phenomena by Cyclotron Resonance Techniques. By DOUSMANIS, G.C. et al. Final Rept. Contract AF 19 604 5479. May 15, 1961. AD 256 598.

DRABBLE, J.R. and J. FENDLEY. The Third Order Elastic Constants of Doped n-Type Germanium. J. PHYS. CHEM. OF SOLIDS, v. 28, Apr. 1967. p. 669-675.

ARMY ELECTRONICS LABS. Dielectric Constant of Germanium as a Function of Frequency, Temperature and Resistivity. by DRUESNE, M.A. TR no. ECOM-2553. Jan. 1965. AD 612 897.

DUDKO, G.V. et al. The Diffusion of Indium and Antimony in Germanium Irradiated with Low Energy Electrons. SOVIET PHYS. SOLID STATE, v. 12, no. 4, Oct. 1970. p. 1016-1017.

DUNLAP, W.C. Jr. Diffusion of Impurities in Germanium. PHYS. REV., v. 94, no. 6, June 1954. p. 1531-1540. [A]

DUNLAP, W.C. Jr. Gold as an Acceptor in Germanium. PHYS. REV., v. 97, no. 3, Feb. 1955. p. 614-629. [B]

DUNLAP, W.C. Jr. Properties of Zinc-, Copper-, and Platinum-Doped Germanium. PHYS. REV., v. 96, no. 1, Oct. 1954. p. 40-45. [C]

ENGELER, W.E. et al. Piezotransmission Measurements of Phonon-Assisted Transitions in Semiconductors. I. Germanium. PHYS. REV., v. 155, no. 3, Mar. 1967. p. 693-702.

FINE, M.E. Elasticity and Thermal Expansion of Germanium between -195 and 275°C. J. OF APPLIED PHYS., v. 24, no. 3, Mar. 1953. p. 338-340.

FISHER, P. and H.Y. FAN. Absorption Spectra and Zeeman Effect of Copper and Zinc Impurities in Germanium. PHYS. REV. LETTERS, v. 5, no. 5, Sept. 1960. p. 195-197.

FLUBACHER, P. et al. The Heat Capacity of Pure Silicon and Germanium and Properties of Their Vibrational Frequency Spectra. PHIL. MAG., v. 4, no. 39, Mar. 1959. p. 273-294.

FREUD, P.J. and G.M. ROTHBERG. Thermoelectric Power of Germanium. Effect of Temperature-Dependent Energy Levels. PHYS. REV., v. 140, no. 3A, Nov. 1965. p. A1007-1014.

FROVA, A. et al. Electro-absorption Effects at the Band Edges of Silicon and Germanium. PHYS. REV., v. 145, no. 2, May 1966. p. 575-583.

FROVA, A. and P. HANDLER. Franz-Keldysh Effect in the Space-Charge Region of a Germanium p-n Junction. PHYS. REV., v. 137, no. 6A, Mar. 1965. p.A1857-A1861.

FROVA, A. and C.M. PENCHINA. Energy Gap Determination in Semiconductors by Electric Field Modulated Optical Absorption. PHYS. STATUS SOLIDI, v. 9, 1965. p. 767-773.

FUHS, W. and J. STUKE. Elastoresistivity of Amorphous Semiconductors. MAT. RES. BULL., v. 5, no. 8, Aug. 1970. p. 611-620.

FUJIYASU, H. et al. Cyclotron Resonance in the Valence Band of Germanium under Uniaxial Stress. J. PHYS. SOC. OF JAPAN, v. 29, no. 3, Sept. 1970. p. 685-695.

FULLER, C.S. et al. Diffusivity and Solubility of Copper in Germanium. PHYS. REV., v. 93, no. 6, Mar. 1954. p. 1182-1189.

FULLER, C.S. and J.A. DITZENBERGER. Effect of Structural Defects in Germanium on the Diffusion and Acceptor Behaviour of Copper. J. OF APPL. PHYS., v. 28, no. 1, Jan. 1957. p. 40-48.

GALLAGHER, C.J. Plastic Deformation of Germanium and Silicon. PHYS. REV., v. 88, no. 4, Nov. 1952. p. 721-722.

GAVINI, A. and M. CARDONA. Modulated Piezoreflectance in Semiconductors. PHYS. REV., B, ser. 3, v. 1, no. 2, Jan. 1970. p. 672-682.

GEBALLE, T.H. and G.W. HULL. Seebeck Effect in Germanium. PHYS. REV., v. 94, no. 5, June 1954. p. 1134-1140.

GEBALLE, T.H. and F.J. MORIN. Ionization Energies of Group III and V Elements in Germanium. PHYS. REV., v. 95, no. 4, Aug. 1954. p. 1085-1086.

GERLICH, D. et al. High Temperature Specific Heats of Germanium, Silicon and Germanium-Silicon Alloys. J. OF APPLIED PHYS., v. 36, no. 1, Jan. 1965. p. 76-79.

GHOSH, A.K. Electroreflectance Spectra and Band Structure of Germanium. PHYS. REV., v. 165, no. 3, Jan. 1968. p. 888-897.

GLASSBRENNER, C.J. and G.A. SLACK. Thermal Conductivity of Silicon and Germanium from 3°K to the Melting Point. PHYS. REV., v. 134, no. 4A, May 1964. p. A1058-A1069.

GLAZOV, V.M. and S.N. CHIZHEVSKAYA. An Investigation of the Magnetic Susceptibility of Germanium, Silicon and Compounds of the Zinc Sulfide Type in the Melting Range and Liquid State. SOVIET PHYS. SOLID STATE, v. 6, no. 6, Dec.1964. p. 1322-1324.

GLAZOV, V.M. et al. Thermal Expansion of Substances Having a Diamondlike Structure and the Volume Changes Accompanying Their Melting. RUSSIAN J. OF PHYS. CHEM., v. 43, no. 2, Feb. 1969. p. 201-205. [A]

GLAZOV, V.M. et al. Temperature Dependence of Thermoelectric Power in Solid and Liquid Germanium. SOVIET PHYS. SEMICONDUCTORS, v. 3, no. 8, Feb. 1970. p. 952-954. [B]

GMELINS HANDBUCH DER ANORGANISCHEN CHEMIE; achte völlig neu bearbeitete Auflage. Germanium. Weinheim, Verlag Chemie, GmbH. 1958.

GOBELI, G.W. and F.G. ALLEN. Photoelectric Properties of Cleaved Gallium Arsenide, Gallium Antimonide, Indium Arsenide and Indium Antimonide Surfaces; Comparison with Silicon and Germanium. PHYS. REV., v. 137, no. 1A Jan. 1965. p. A245-A254.

GOLIKOVA, O.A. and A.V. PETROV. Electron Mobility in Germanium in the Temperature Range 300-1000°K. SOVIET PHYS. SOLID STATE, v. 6, no. 10, Apr. 1965. p. 2443-2446.

GORYUNOVA, N.A. The Chemistry of Diamondlike Semiconductors. Ed. J.C. ANDERSON, The M.I.T. Press, Mass. Inst. of Tech., Cambridge, Mass. 1963, 236 p.

GOUSKOV, L. et al. Carrier Recombination in Nickel-Doped Germanium from Lifetime and Noise Measurements. PHYS. STATUS SOLIDI, v. 21, no. 2, June 1967. p. 619-626.

GRIECO, A. and H.C. MONTGOMERY. Thermal Conductivity of Germanium. PHYS. REV., v. 86, no. 4, May 1952. p. 570.

GRIGOROVICI, R. et al. Thermoelectric Power in Amorphous Germanium. PHYS. STATUS SOLIDI, v. 16, no. 2, 1966. p. K143-K145.

GROVES, S.H. et al. Infrared Magnetoelectroreflectance in Germanium, Gallium Antimonide and Indium Antimonide. PHYS REV. LETTERS, v. 17, no. 12, Sept. 1966. p. 643-646.

GUNDJIAN, A.A. Electrostriction in Germanium. SOLID STATE COMM., v. 3, no. 9, Sept. 1965. p. 279-281.

HAAS, C. Infrared Absorption in Heavily Doped n-Type Germanium. PHYS. REV., v. 125, no. 6, Mar. 1962. p. 1965-1971.

HALL, R.N. and J.H. RACETTE. Diffusion and Solubility of Copper in Extrinsic and Intrinsic Germanium, Silicon and Gallium Arsenide. J. APPL. PHYS. v. 35, no. 2, Feb. 1964. p. 379-397.

HALPERN, J. and B. LAX. Magnetoabsorption of the Indirect Transition in Germanium. J. OF PHYS. AND CHEM. OF SOLIDS, v. 26, no. 5, May 1965. p. 911-919.

HARBEKE, G. Intrinsic Optical Absorption in Germanium. PHYS. STATUS SOLIDI, v. 5, no. 3, 1964. p. 548-552.

HENSEL, J.C. Microwave Combined Resonances in Germanium; g-Factor of the Free Hole. PHYS. REV. LETTERS, v. 21, no. 14, Sept. 1968. p. 983-986.

HERGENROTHER, K.M. and J.M. FELDMAN. Radiative Recombination in Tin-Doped Germanium. J. OF APPL. PHYS., v. 40, no. 5, Apr. 1969. p. 2323-2324.

HERRING, C. et al. Phonon-Drag Thermomagnetic Effects in n-Type Germanium. I. General Survey. PHYS. REV., v. 111, no. 1, July 1958. p. 36-37.

HIGGINBOTHAM, C.W. et al. Intrinsic Piezobirefringence of Germanium, Silicon and Gallium Arsenide. PHYS. REV., v. 184, no. 3, Aug. 1969. p. 821-829.

HOBDEN, M.V. Direct Optical Transitions from the Split-Off Valence Band to the Conduction Band in Germanium. J. OF PHYS. AND CHEM. OF SOLIDS, v. 23, June 1962. p. 821-822.

HUI, W.L.C. and J.P. CHOPRA. Seebeck Coefficient of Thin-Film Germanium. J. OF APPL. PHYS., v. 38, no. 9, Aug. 1967. p. 3477-3478.

HUNTER, W.R. et al. Thin Films and Their Uses for the Extreme Ultraviolet. APPLIED OPTICS, v. 4, no. 8, Aug. 1965. p. 891-898.

IGLITSYN, M.U. and E.S. YUROVA. Recombination of Nonequilibrium Charge Carriers at Cadmium Ions in Germanium. SOVIET PHYS. SOLID STATE, v. 7, no. 3, Sept. 1965. p. 723-724.

IGNATKOV, V.D. and V.E. KOSENKO. Diffusion of Tellurium in Germanium. SOVIET PHYS. SOLID STATE, v. 4, no. 6, Dec. 1962. p. 1193-1196.

IVANOV, Yu. L. Temperature Dependence of the Activation Energy of the Second and Third Levels of Copper in Germanium. SOVIET PHYS. SOLID STATE, v. 5, no. 4, Oct. 1963. p. 888-889.

JAEGER, H. and W. KOSAK. The Metal-Semiconductor-Contact Barriers of Metals from the First and Eighth Transition Groups, on Silicon and Germanium (In Ger.). SOLID STATE ELECTRONICS, v. 12, no. 7, July 1969. p. 511-518.

JAYARAMAN, A. et al. Delta-1 Conduction Band Minimum of Germanium from High Pressure Studies on p-n Junctions. PHYS. REV., v. 171, no. 3, July 1968. p. 836-838.

JONES, R.L. and P. FISHER. Excitation Spectra of Group III Impurities in Germanium. J. OF PHYS.AND CHEM. OF SOLIDS, v. 26, no. 7, July 1965. p. 1125-1131. [A]

JONES, R.L. and P. FISHER. Spectroscopic Study of the Deformation Potential Constants of Group III Acceptors in Germanium. PHYS. REV., B, v. 2, no. 6, Sept. 1970. p. 2016-2029. [B]

KAISER, W. et al. Infrared Absorption and Oxygen Content in Silicon and Germanium. PHYS. REV., v. 101, no. 4, Feb. 1956. p. 1264-1268.

KASPER, J.S. and S.M. RICHARDS. The Crystal Structures of New Forms of Silicon and Germanium. ACTA CRYSTALLO-GRAPHICA, v. 17, pt. 6, June 1964. p. 752-755.

KLINGER, Y. The Conductivity of Germanium at 2.4×10^{10} Hz. PHYS. REV., v. 92, no. 2, Oct. 1953. p. 509-510.

KLINKOVA, O.A. and O.R. NUYAZOVA. Radiation Accelerated Diffusion of Gold in Silicon. SOVIET PHYS. SOLID STATE, v. 12, no. 7, Jan. 1971. p. 1760-1761.

KOENIG, S.H. Recombination of Thermal Electrons in n-Type Germanium below 10°K. PHYS. REV., v. 110, no. 4, May 1958. p. 988-990.

KOENIG, S.H. et al. Electrical Conduction in n-Type Germanium at Low Temperatures. PHYS. REV., v. 128, no. 4, Nov. 1962. p. 1210-1214.

KOSENKO, V.E. Diffusion and Solubility of Cadmium in Germanium. SOVIET PHYS. SOLID STATE, v. 1, no. 10, Apr. 1960. p. 1481-1484.

KUROVA, I.A. and S.G. KALASHNIKOV. Electrical Conductivity of Germanium at Low Temperatures. SOVIET PHYS. SOLID STATE, v. 1, no. 9, Sept. 1959. p. 1353-1356. [A]

KUROVA, I.A. and S.G. KALASHNIKOV. Ionization Energies of Bismuth and Thallium in Germanium. SOVIET PHYS. TECH. PHYS., v. 3, no. 2, Feb. 1958. p. 232-234. [B]

KUROVA, I.A. et al. Impurity Photoconductivity Spectra of p-Type Germanium with Gallium, Mercury, Gold and Nickel Impurities. SOVIET PHYS. JETP, v. 24, no. 2, Feb. 1967. p. 268-271.

LaVILLA, R.E. and H. MENDLOWITZ. Optical Properties of Germanium. J. OF APPL. PHYS., v. 40, no. 8, July 1969. p. 3297-3300.

LEADBETTER, A.J. and G.R. SETTATREE. Anharmonic Effects in the Thermodynamic Properties of Solids. VI. Germanium Heat Capacity between 30 and 500°C and Analysis of Data. J. PHYS., C. (SOLID STATE PHYS.), v. 2, ser. 2, 1969. p. 1105-1111.

LETAW, H. Jr. et al. Self Diffusion in Germanium. PHYS. REV., v. 102, no. 3, May 1956. p. 636-639.

LEVINGER, B.W. and D.R. FRANKL. Cyclotron Resonance Measurements of the Energy Bank Parameters of Germanium. J. OF PHYS. AND CHEM. OF SOLIDS, v. 20, no. 3/4, Aug. 1961. p. 281-288.

LINSTEADT, G. Infrared Transmittance of Optical Materials at Low Temperatures. APPLIED OPTICS, v. 3, no. 12, Dec. 1964. p. 1453-1456.

LUKES, F. On the Theory of the Temperature Dependence of the Refractive Index of Insulators and Semiconductors. CZECH. J. OF PHYS., v. 8, no. 4, 1958. p. 423-434.

LUKES, F. and E. SCHMIDT. Electroreflectance Measurements with the Electrolyte Method in the Infrared. PHYS. LETTERS, v. 23, no. 7, Nov. 1966. p. 413-414.

MacFARLANE, G.G. et al. Exciton and Phonon Effects in the Absorption Spectra of Germanium and Silicon. PHYS. AND CHEM. OF SOLIDS, v. 8, Jan. 1959. p. 388-392. [A]

MacFARLANE, G.G. et al. Fine Structure in the Absorption Edge Spectrum of Germanium. PHYS. REV., v. 108, no. 6, Dec. 1957. p. 1377-1383. [B]

McSKIMIN, H.J. Measurement of Elastic Constants at Low Temperatures by Means of Ultrasonic Waves; Data for Silicon and Germanium Single Crystals and for Fused Silica. J. OF APPL. PHYS., v. 24, no. 8, Aug. 1953. p.988-997.

McSKIMIN, H.J. and P. ANDREATCH, Jr. Elastic Moduli of Germanium vs. Hydrostatic Pressure at 25.0°C and -195.8°C. J. OF APPL. PHYS., v. 34, no. 3, Mar. 1963. p. 651-655. [A]

McSKIMIN, H.J. and P. ANDREATCH, Jr. Measurement of Third Order Moduli of Silicon and Germanium. J. OF APPL. PHYS., v. 35, no. 11, Nov. 1964. p. 3312-3319. [B]

MEER, W. and D. POMMERRENIG. Diffusion of Aluminum and Boron in Germanium (In Ger.). ZEITSCHRIFT FUER ANGEWANDTE PHYSIK, v. 23, no. 6, 1967. p. 369-372.

MILLER, R.C. and F.M. SMITS. Diffusion of Antimony out of Germanium and Some Properties of the Antimony-Germanium System. PHYS. REV., v. 107, no. 1, July 1957. p. 65-70.

MARTON, L. and J. TOOTS. Optical Properties of Germanium in the Far Ultraviolet. PHYS. REV., v. 160, no. 3, Aug. 1967. p. 602-606.

MARUCCHI, J. Mobility Variation in Thin Germanium Films, Oriented by Epitaxial Deposition. (In Fr.) ACAD. DES SCI., C.R., v. 260, no. 25, June 1965. p. 6580-6582.

MINOMURA, S. and H.G. DRICKAMER. Pressure induced Phase Transitions in Silicon, Germanium and Some III-V Compounds. J. OF PHYS. AND CHEM. OF SOLIDS, v. 23, May 1962. p. 451-456.

MORIN, F.J. Lattice Scattering Mobility in Germanium. PHYS. REV., v. 93, no. 1, Jan. 1954. p. 62-63.

MORIN, F.J. and J.P. MAITA. Conductivity and Hall Effect in the Intrinsic Range of Germanium. PHYS. REV., v. 94, no. 6, June 1954. p. 1525-1529.

MURASE, K. et al. Determination of Deformation Potential Constants from the Electron Cyclotron Resonance in Germanium and Silicon. PHYS. SOC. JAPAN, J., v. 29, no. 5, Nov. 1970. p. 1248-1257.

NISHINO, T. and Y. HAMAKAWA. Low Temperature Electroreflectance Spectra of Germanium in Spectral Region from 0.7 to 2.6 eV. PHYS. SOC. JAPAN, J., v. 26, no. 2, Feb. 1969. p. 403-412.

OSTROBORODOVA, V.V. Degeneracy Factors of Impurity Levels and the Analysis of Electrical Properties of Germanium Doped with Gold. SOVIET PHYS. SOLID STATE, v. 7, no. 2, Aug. 1965. p. 484-490.

PAIGE, E.G.S. The Drift Mobility of Electrons and Holes in Germanium at Low Temperatures. PHYS. AND CHEM. OF SOLIDS, v. 16, no. 3/4, Jan. 1961. p. 207-219.

PAJASOVA, L. Reflection Spectrum of Germanium in Ultraviolet Region. SOLID STATE COMM., v. 4, no. 11, Nov. 1966. p. 619-620.

PALMBERG, P.W. Secondary Emission Studies on Germanium and Sodium-Covered Germanium. J. OF APPL. PHYS., v. 38, no. 5, Apr. 1967. p. 2137-2147.

PANKOVE, J.I. A Contribution to the Study of Infrared Emission in Germanium (In Fr.). ANNALES DE PHYSIQUE, v. 6, no. 3/4, Mar./Apr. 1961. p. 331-374.

PANTELEEV, V.A. Diffusion of Indium in Germanium along Dislocations. SOVIET PHYS. SOLID STATE, v. 7, no. 3, Sept. 1965. p. 734-736.

PARKER, J.H. et al. Raman Scattering by Silicon and Germanium. PHYS. REV., v. 155, no. 3, Mar. 1967. p. 712-714.

PAUL, W. Band Structure of the Intermetallic Semiconductors from Pressure Experiments. J. OF APPL. PHYS., Supp. to v. 32, no. 10, Oct. 1961. p. 2083-2094.

PAUL, W. and H. BROOKS. Effect of Pressure on the Properties of Germanium and Silicon. In PROGRESS IN SEMI-CONDUCTORS. Ed. by GIBSON, A. and R. BURGESS. London, Heywood and Co., 1963. v. 7. p. 135-238.

PIESBERGEN, U. The Mean Atomic Heats of the III-V Semiconductors, AlSb, GaAs, InP, GaSb, InAs, InSb and the Atomic Heats of the Element Germanium between 12 and 273°K (In Ger.). ZEITSCHRIFT FUER NATURFORSCHUNG, v. 18a, no. 2, p. 141-147, Feb. 1963.

PILLER, H. and V.A. PATTON. Interband Faraday Effect in Aluminum Antimonide, Germanium and Gallium Antimonide. PHYS. REV., v. 129, no. 3, Feb. 1963. p. 1169-1173.

PÜDÖR, B. Electron Mobility in Plastically Deformed High Purity Germanium. PHYS. STATUS SOLIDI, v. 16, no. 2, 1966. p. K167-K170.

POLLAK, F.H. and M. CARDONA. Piezo-Electroreflectance in Germanium, Gallium Arsenide and Silicon. PHYS. REV., v. 172, no. 3, Aug. 1968. p. 816-837.

POTTER, R.F. Optical Constants of Germanium in Spectral Region from 0.5 to 3.0 eV. PHYS. REV., v. 150, no. 2, Oct. 1966. p. 562-567.

PRATT, B. and F. FRIEDMAN. Diffusion of Lithium into Germanium and Silicon. J. APPL. PHYS., v. 37, no. 4, Mar. 1966, p. 1893-1896.

RANK, D.H. et al. The Index of Refraction of Germanium Measured by an Interference Method. OPTICAL SOC. OF AMERICA, J., v. 44, no. 1, Jan. 1954. p. 13-16.

REMENYUK, A.D. et al. Forced Birefringence in n-Type Germanium. SOVIET PHYS. SEMICONDUCTORS, v. 1, no. 7, Jan. 1968. p. 934-936.

RISKAER, S. Determination of Shear Deformation from the Free Carrier Piezobirefringence in Germanium and Silicon. PHYS. REV., v. 152, no. 2, Dec. 1966. p. 845-849.

ROLLIN, B.V. and J.M. ROWELL. Hot Carriers in Germanium. PHYS. SOC., PROC., v. 76, pt. 6, Dec. 1960, p. 1001-1002.

ROUZEYRE, M. et al. Electro-Magneto-Reflectivity in Germanium at 1.5°K (In Fr.). SOLID STATE COMM., v. 7, no. 17, Sept. 1969. p. 1219-1223.

SALZBERG, C.D. and J.J. VILLA. Index of Refraction of Germanium. OPTICAL SOC. OF AMERICA, J., v. 48, no. 8, Aug. 1958. p. 579.

SCHETZINA, J.F. and J.P. McKELVY. Strain Dependence of the Minority Carrier Mobility in p-Type Germanium. PHYS. REV., v. 181, no. 3, May 1969. p. 1191-1195.

SCHMIDT-TIEDEMANN, K.J. Stress Optical Constants of Germanium. J. APPL. PHYS., v. 32, no. 10, Oct. 1961, p. 2058-2059.

SERAPHIN, B.O. et al. Field Effect of the Reflectivity in Germanium. J. APPL. PHYS., v. 36, no. 7, July 1965. p. 2242-2250.

SHARMA, B.L. Diffusion in Semiconductors. TRANS. TECH. PUBL. Adolf Ey Str. 1d D3392. Clausthal Zellerfeld, Gy.

SHENKER, H. et al. Photoconductive and Electrical Properties of Uncompensated Beryllium-Doped Germanium. AIME, METALL. SOC., TRANS., v. 239, no. 3, Mar. 1967. p. 347-349.

SIDOROV, V.I. and T.M. LIFSHITS. Photoconductivity in Germanium Doped with Group III Impurities Caused by Optical Excitation of the Impurity Level. SOVIET PHYS. SOLID STATE, v. 8, no. 8, Feb. 1967. p. 2000-2002.

SIDOROV, V.I. et al. Investigation of Optical Absorption in Germanium Doped with Zinc and Compensated with Antimony. SOVIET PHYS. SOLID STATE, v. 8, no. 7, Jan. 1967. p. 1608-1610.

SILFVAST, W.T. et al. Laser Action in Singly Ionized Germanium, Tin, Lead, Indium, Cadmium and Zinc. APPL. PHYS. LETTERS, v. 8, no. 12, June 1966. p. 318-319.

SLATTER, J.A.G. Radiative Electron Capture by Gold and Platinum Impurities in Germanium. PHYS. STATUS SOLIDI, v. 40, no. 1, July 1970. p. 31-41.

SLOAN, B.J. and J.R. HAUSER. Effects of Uniaxial Compressive Stress on Minority Carrier Lifetime in Silicon and Germanium. J. APPL. PHYS., v. 41, no. 8, July 1970. p. 3504-3508.

SMITH, C.S. Piezoresistance Effect in Germanium and Silicon. PHYS. REV., v. 94, no. 1, Apr. 1954. p. 42-49.

SPARKS, B.W. and C.A. SWENSON. Thermal Expansion from 2 to 40°K of Germanium, Silicon and Four III-V Compounds. PHYS. REV., v. 163, no. 3, Nov. 1967. p. 779-790.

TAGIROV, V.I. and A.A. KULIEV. Diffusivity and Solubility of Thallium in Germanium. SOVIET PHYS. SOLID STATE, v. 4, no. 1, July 1962. p. 196-198.

TYAPKINA, N.D. Impurity Photoconductivity of Beryllium Doped p-Type Germanium. SOVIET PHYS. SEMICONDUCTORS, v. 3, no. 2, Aug. 1969. p. 178-181.

TYLER, W.W. Properties of Silver-Doped Germanium. AMER. PHYS. SOC., BULL., v. 3, 1958. p. 128. [A]

TYLER, W.W. Deep Level Impurities in Germanium. PHYS. AND CHEM. OF SOLIDS, v. 8, 1959. p. 59-65. [B]

TYLER, W.W. et al. Properties of Germanium Doped with Nickel. PHYS. REV., v. 98, no. 2, Apr. 1955. p. 461-465. [A]

TYLER, W.W. et al. Properties of Germanium Doped with Cobalt. PHYS. REV., v. 97, no. 3, Feb. 1955, p. 669-672.
   [B]

TYLER, W.W. and H.H. WOODBURY. Properties of Germanium Doped with Iron. I. Electrical Conductivity. PHYS. REV.,
v. 96, no. 4, Nov. 1954. p. 874-882.

UKHANOV, Yu. I. and Yu. V. MALTSEV. Investigation of the Temperature Dependence of the Electron Effective Mass in
Semiconductors. SOVIET PHYS. SOLID STATE, v. 5, no. 10, Apr. 1964. p. 2144-2149.

VAN DER MAESEN, F. and J.A. BRENKMAN. The Solid Solubility and the Diffusion of Nickel in Germanium. PHILIPS RES.
REPORTS, v. 9, 1954. p. 225-230.

WALLEY, P.A. and A.K. JONSCHER. Electrical Conduction in Amorphous Germanium. THIN SOLID FILMS, v. 1, no. 5,
Mar. 1968. p. 367-377.

van WIERINGEN, A. and N. WARMOLTZ. On the Permeation of Hydrogen and Helium in Single Crystal Silicon and
Germanium at Elevated Temperatures. PHYSICA, v. 22, 1956. p. 849-865.

WALES, J. et al. Optical Properties of Germanium Films in the 1 to 5 Microns Range. THIN SOLID FILMS, v. 1,
no. 2, Sept. 1967. p. 137-150.

WEBER, R.E. and W.T. PERIA. Work Function and Structural Studies of Alkali-Covered Semiconductors. SURFACE SCI.,
v. 14, no. 1, 1969. p. 13-38.

WEI, L.Y. Diffusion of Silver, Cobalt and Iron in Germanium. PHYS. CHEM. SOLIDS, v. 18, no. 2/3, 1967, p. 162-174.

WHITE, G.K. and S.B. WOODS. Thermal Conductivity of Germanium and Silicon at Low Temperatures. PHYS. REV., v. 103,
no. 3, Aug. 1956. p. 569-571.

WOLFF, G.A. et al. Relationship of Hardness, Energy Gap and Melting Point of Diamond Type and Related Structures.
In SEMICONDUCTORS AND PHOSPHORS, PROC., Internat. Colloquium, 1956, Garmisch-Partenkirchen. Ed. by SCHON, M. and
H. WELKER. N.Y. Interscience, 1958. p. 463-469.

WÖLFLE, R.and J. DORENDORF. The Determination of the Diffusion Coefficients of Arsenic in p-n Junction Germanium.
(In Ger.) SOLID STATE ELECTRONICS, v. 5, Mar./Apr. 1962. p. 98-102.

WOODBURY, H.H. and W.W. TYLER. Properties of Germanium Doped with Manganese. PHYS. REV., v. 100, no. 2, Oct.
1955. p. 659-662. [A]

WOODBURY, H.H. and W.W. TYLER. Triple Acceptors in Germanium. PHYS. REV., v. 105, no. 1, Jan. 1957. p. 84-92.
   [B]

YU, P.Y. and M. CARDONA. Temperature Coefficient of the Refractive Index of Diamond- and Zincblende-Type Semiconductors. PHYS. REV., B, ser. 3, v. 2, no. 8, Oct. 1970. p. 3193-3198.

ZALLEN, R. and W. PAUL. Effect of Pressure on Interband Reflectivity Spectra of Germanium and Related Semiconductors. PHYS. REV., v. 155, no. 3, Mar. 1967. p. 703-711.

ZAVARITSKII, N.V. Shift of Singularities of Germanium Lattice Vibration Spectrum under Pressure. JETP LETTERS,
v. 12, no. 1, July 1970. p. 18-20.

ZEMSKII, V.I. and I.V. MOCHAN. Investigation of Indirect Transitions in Germanium by the Electrical Absorption
Method. SOVIET PHYS. SOLID STATE, v. 11, no. 9, Mar. 1970. p. 2124-2128.

ZWERDLING, S. et al. Exciton and Magneto-Absorption of the Direct and Indirect Transitions in Germanium. PHYS.
REV., v. 114, no. 1, Apr. 1959. p. 80-90.

PHYSICAL PROPERTIES	SYMBOL	VALUE	UNIT	NOTES	TEMP.($^{\circ}$K)	REFERENCES
Formula		Si				
Atomic Weight		28.09				
Density		2.32831	g/cm^3		291	Donnay
		2.30		solid at melting point	1410°C	Glazov et al.
		2.53		liquid	1420°C	
Color						
Bulk Single Crystal		steel-grey		shiny, opaque		Gmelin, p. 16
Film		light-yellow		translucent		
Hardness		7	Mohs			Gmelin, p. 55
Knoop Microhardness	H_{25}	1150	kg/mm^2			Wolff et al.
Cleavage		(111)		good		Goryunova, p. 66
Symmetry		cubic, diamond				Donnay
Space Group		Fd3m Z-2				Wyckoff
Lattice Parameter	a_o	5.43089	$\overset{\circ}{A}$		300	Kiendl
	Si-Si	2.351				Gmelin, p. 23
Atomic Volume		12.057	$cm^3/gr\ atom$		300	Gmelin, p. 26
Melting Point		1420	$^{\circ}$C			Glazov et al.
Pressure Coeff.		-5.8	$^{\circ}$C/kbar			Jayaraman
Boiling Point		2787	$^{\circ}$C			Honig
Heat of Sublimation		4.91	kcal/gr		298	Honig
Vapor Pressure		10^{-5}	mm Hg		1177°C	Honig
		10^{-7}			777°C	
Heat of Fusion		430	cal/gr			Gmelin, p. 69
Entropy		4.497	cal/gr atom $^{\circ}$K			Flubacher et al.
Zero Point Energy		1435	cal/gr atom		298	Flubacher et al.
Liquid Surface Tension		720	$ergs/cm^2$		1415°C	Zadumkin
Temperature Coeff.		-0.072	$ergs/cm^2\ ^{\circ}$K			
Specific Heat		0.171	cal/gr $^{\circ}$K	high purity, single crystal	300	Flubacher et al.
		0.165		high purity, single crystal	0°C	Shanks et al.
		0.184			100°C	
		0.2215			800°C	
Thermal Diffusivity		0.85	cm^2/sec		27°C	Shanks et al.
Debye Temperature		645	$^{\circ}$K	specific heat meas.	0	Flubacher et al., Keesom & Seidel
		674		thermal conductivity meas.	300	Glassbrenner & Slack

SILICON

PHYSICAL PROPERTIES	SYMBOL	VALUE	UNIT	NOTES	TEMP.(°K)	REFERENCES
Thermal Conductivity		10	W/cm °K	high purity, polished	4	Hurst & Frankl
		26		high purity	50	Glassbrenner &
		9.5			100	Slack
		7.52		single crystals	100	Fulkerson et al.
		1.40			300	
		0.736			500	
		0.306			1000	
		0.239			1350	
		0.287			1400	Shanks et al.
		0.216			1680	Glassbrenner & Slack
Thermal Expansion Coeff.		0.0056	10^{-8}/°K		4	Sparks & Swenson
		0.0032	10^{-6}/°K		16	
		-0.0475			32	
		-0.135	10^{-6}/°K		40	Gibbons
		-0.34			100	Carr et al., Gibbons
		-0.04			120	Carr et al.
		+0.135			130	
		0.52			150	Carr et al., Ibach
		1.44			200	
		2.44			300	
		4.08			850	Ibach, Maissel
Elastic Coefficient						
Stiffness	c_{11}	16.48	10^{11}dyne/cm^2	Debye-Sears effect on single crystals, p-type, high purity, resistivity of 160 ohm-cm	298	Metzger & Kessler, Ezz-El-Arab et al.
	c_{12}	6.35				
	c_{44}	7.90				
	c_{11}	16.77		high purity crystals, 150 ohm-cm resistivity	77	McSkimin & Andreatch A
	c_{12}	6.50				
	c_{44}	8.04		high arsenic doping	78,300	Beilin et al.
Temp. Coeff.	$(1/c_{11})(\Delta c_{11}/\Delta T)$	-122	10^{-6}/°K		300	Metzger & Kessler
	$(1/c_{12})(\Delta c_{12}/\Delta T)$	-162				
	$(1/c_{44})(\Delta c_{44}/\Delta T)$	- 97				

		4°K	77°K				
Pressure Coeff.	$\Delta c_{11}/\Delta P$	4.19	4.25	high purity crystals, 150 ohm-cm resistivity		Beattie & Schirber, McSkimin & Andreatch A
	$\Delta c_{12}/\Delta P$	4.02	4.11			
	$\Delta c_{44}/\Delta P$	0.80	0.80			

PHYSICAL PROPERTIES	SYMBOL	VALUE	UNIT	NOTES	TEMP.(°K)	REFERENCES
Compliance	s_{11}	0.077	10^{-11}cm^2/dyne	calc.	300	Landolt-Börnstein
	s_{12}	-0.021				
	s_{44}	0.126				
Bulk Modulus K		9.788	10^{11}dyne/cm^2	P=2109 kg/cm^2	300	McSkimin & Andreatch A
Volume Compressibility 1/K		0.102	10^{-11}cm^2/dyne		300	McSkimin & Andreatch A
		0.28		liquid	1793	Baidov & Gitis

SILICON

PHYSICAL PROPERTIES	SYMBOL	VALUE		UNIT	NOTES	TEMP.($^\circ$K)	REFERENCES
Third Order Moduli	c_{111}	-82.5		10^{11}dyne/cm^2	ultrasonic wave meas.	298	McSkimin & Andreatch B
	c_{112}	-45.1					
	c_{123}	- 6.4					
	c_{144}	+ 1.2					
	c_{166}	-31.0					
	c_{456}	- 6.4					
Young's Modulus		16.9		10^{11}dyne/cm^2	in [111] plane	298	Wortman & Evans
Shear Modulus		6.7		10^{11}dyne/cm^2	in [111] plane	298	Wortman & Evans
Poisson's Ratio		0.358			[111] calc.	298	Wortman & Evans
		0.279			[100] calc.		
		0.180					Metals Handbook
Sound Velocity	(110)*+	(100)*					
Longitudinal	9101	8396		m/sec	30 and 57 MHz	298	*Metzger & Kessler,
Transverse$_1$	5824	5808					+Ess-Al-Arab et al.
Transverse$_2$	4662						
		3920		m/sec	liquid	1793	Baidov & Gitis

ELECTRICAL PROPERTIES

Dielectric Constant

	SYMBOL	VALUE	UNIT	NOTES	TEMP.($^\circ$K)	REFERENCES
Static	ε_o	12.1		capacitance meas.	4.2	Rao & Smakula
		11.9		at 7500 MHz	300	
		11.7		1 MHz	77	Dunlap & Watters
Pressure Coeff. $(1/\varepsilon)(d\varepsilon/dP)$		-4	10^{-7}/kg cm^{-2}	capacitance meas. 10 MHz	263	Cardona et al. B
Temp. Coeff. $(1/\varepsilon)(d\varepsilon/dT)$		7.8	10^{-5}/$^\circ$K		77-200	
Electrical Resistivity ρ		1	10^8 ohm-cm	high purity p-type	20	Finke & Lautz
		2.3	10^5	high purity p-type, single crystal	300	Hoffman et al.
		1	10^2		475	
Temp. Coeff. $(1/\rho)(d\rho/dT)$		-8.1	10^{-2}/$^\circ$K	high purity p-type, T=273-323°K	300	Bullis et al.
		-9.4			273	
Pressure Coeff. $(1/\rho)(d\rho/dP)$		-1.4	10^{-6}/kg cm^{-2}	P=(25-100)10^3 kg/cm^2	300	Vereshchagin et al.
		-0.9		P=(36-100)10^3 kg/cm^2	300	Bridgman
Electrical Resistivity		$\sim 10^6$	ohm-cm	amorphous films, 0.3-10µ thick, in vacuo	300	Brodsky et al.
		$\sim 10^{10}$	ohm-cm	amorphous films, 0.2-4µ thick	300	Chittick et al.
	$d\rho/dT$	-83	ohm-cm /$^\circ$K		20-140	
Mobility						
Electron	μ_n	100,000	cm^2/V sec	$n_n= 10^{14}$	20	Logan & Peters
Hole	μ_p	80,000		$n_p= 10^{14}$		

SILICON

ELECTRICAL PROPERTIES	SYMBOL	VALUE		UNIT	NOTES	TEMP.($^\circ$K)	REFERENCES
Mobility							
Electron	μ_n	10,000		cm^2/V sec	pure p-type	50	Finke & Lautz
		1,000				200	
	μ_n	1,880			pure, $n_n < 10^{14}$, 800 ohm-cm	300	Messier & Flores
Hole	μ_p	398			pure, $n_p < 10^{14}$, 8000 ohm-cm		
	μ_n	400			1.7μ thick epitaxial film	300	Dumin & Ross
	μ_p	160					
	μ_n	10,000			epitaxial Si film on	77	Schloetterer
		2,200			Si substrate, $n_n = 10^{16}$, low	200	
		1,100			crystal defect density	300	
Drift Mobility							
Electron	μ_n	80,000				30	Joergensen et al.
		20,000				77	
		100°K	300°K				
	μ_n	10,000	1,350		in p-type silicon		Ludwig & Watters
	μ_p	80,000	500		in n-type silicon; both types are high purity		
Drift Velocity					Field (V/cm)		
Electron		12		10^6cm/sec	3×10^3	4	Cecchi et al.
Hole		8					
Electron		10			2×10^4	77	Su et al.
		8				300	
Hole		3			9×10^3	300	Quaranta et al.
Mobility							
Temperature Coeff.							
Electron		$T^{-2.06}$			n-type	100-320	Messier & Flores
Hole		$T^{-2.91}$			p-type		
Hole		$T^{+1.5}$			p-type	4-60	Finke & Lautz
		$T^{-1.5}$				80-125	
Drift Mobility							
Temperature Coeff.							
Electron		$T^{-2.5}$			n-type	160-400	Ludwig & Watters
		$T^{-2.1}$			n-type	500-1670	Burton & Madjid
Hole		$T^{-2.7}$			p-type	160-400	Ludwig & Watters
Lifetime							
Electron		130		μsec	high purity p-type, dislocation-free	300	Noack
		3			high purity n-type	50	Leadon & Naber
		500				300	
Hole		100			p-type	300	Mattis et al.

SILICON

ELECTRICAL PROPERTIES	SYMBOL	VALUE			UNIT	NOTES	TEMP.(°K)	REFERENCES
			°K					
Piezoresistance Coeff.		300	77	4-10				
	π_{11}	-102	-350	-90	10^{-12} cm^2/dyne	n-type		Smith (300°K), Tufte & Stelzer (77°K), Sasaki & Kinoshita (4-10°K)
	π_{12}	-53.4						
	π_{44}	-13.6						
	π_{11}	15	6.6			p-type		Smith (77°K)
	π_{12}		-1.1					Landwehr (300°K)
	π_{44}		138.1					

Elastoresistance Coeff.		n-type	p-type				300	Smith
	m_{11}	-71.0	12.9					
	m_{12}	88	5.1					
	m_{44}	-10.8	110.0					

Elastoresistivity $1/\epsilon(\Delta\rho/\rho)$								
		-5					500	Fuhs & Stuke
		-1					300	
		+5					125	

Effective Mass								
Electron	m_n	1.18					300	Barber
Longitudinal	$m_{n_\parallel}$	0.9163			m_o	high purity n-type	1.26	Hensel
Transverse	$m_{n_\perp}$	0.1905						
Hole		0.81					300	Barber
Heavy Hole	m_{hp}	0.59				p-type	4	Dresselhaus et al.
Light Hole	m_{lp}	0.16						
Density of States	m_{dn}	1.062					4	Barber
	m_{dp}	0.591						
Split-off Valence Band	m_{so}	0.29				cyclotron resonance with uniaxial stress	4.25	Ohyama et al.

Diffusion Coeff. and Energy Levels	Dopant	D_o	D	E_{act}	E_a	E_d	Meas.	Temp.(°C)	
		(cm^2/sec)			(eV)				
	Ag	2.0×10^{-3}		1.6				1100-1350	Boltaks & Shih Yin
			3×10^{-9}					1100	
					0.29 CB	0.26 VB	elec.	20	Thiel & Gandhi
	Al	4.8		3.3				1050-1450	Miller & Savage
			2×10^{-12}					1100	
					0.057		elec.		Holland & Paul
					0.067		opt.		Burstein et al. B
	Al-Zn				0.078		elec.		Fuller & Morin
	As	60		4.2				850-1150	Masters & Fairfield A
			10^{-14}					1100	

ELECTRICAL PROPERTIES

REFERENCES

Diffusion Coeff. and Energy Levels

Dopant	D_o	D	E_{act}	E_a	E_d	Meas.	Temp.(°C)	
				(cm²/sec)	(eV)			
As	0.32		3.6				1050-1350	Fuller & Ditzenberger
		3×10^{-14}					1100	
				0.049		elec.		Morin & Maita, Holland & Paul
				0.0533		opt.		Picus et al., Bichard & Giles
Au	1.1×10^{-3}		1.1				800-1300	Struthers
		4×10^{-8}					1100	
		5×10^{-5}				dislocation-free	1000	Badalov & Shurman
				0.54 CB	0.77 CB	photovoltaic λ<1.5μ	55-80	Tasch & Sah
				0.34		photocond.	90°K	Glinchuk et al.
				0.54 CB	0.35 VB	elec. & opt.	300°K	Boltaks et al., Collins et al., Nathan & Paul
		10^{-12}				neutron irradiated Si	800	Zyuz et al.
Au-Zn				0.55		photocond.	90°K	Glinchuk et al.
B	16		3.8			electrolum.	1050-1350	Williams
		4×10^{-13}					1100	
		2×10^{-12}	4.0					Uskov et al., Sladkov et al.
				0.045		elec. & opt.		Morin & Maita, Fuller & Morin, Burstein et al. A & B
				0.084 0.052		elec.		Tetelbaum
B-Zn				0.092 0.126		elec.		Fuller & Morin
Be	$\sim 10^{-7}$			0.17		elec.	1050	Taft & Carlson
C	0.33		2.92				1070-1400	Newman & Wakefield
	10^{-11}						1100	
Cd	$\sim 10^{-8}$						1200	Collins & Carlson
	10^{-6}-10^{-4}			0.10 VB 0.30		elec. & opt.	1000-1300	Bakhadyrkhanov B & C
				0.22 CB 0.37 0.52				
Co	$\sim 10^{-3}$			0.37 VB 0.39 CB		elec.	900-1100	Collins & Carlson, Irmler

ELECTRICAL PROPERTIES

Diffusion Coeff. and Energy Levels	Dopant	D_o	D (Values)	E_{act}	E_a	E_d	Meas.	Temp.(°C)	
		(cm^2/sec)			(eV)				
	Co			0.35 VB 0.53 CB			elec. & opt.		Penchina et al.
				0.53 CB	0.40 VB		elec.	220-300°K	Moore et al.
				0.44			elec.		Ghandi et al.
	Cr		$\sim 10^{-8}$					1200	Collins & Carlson
	Cu	4×10^{-2}		1.0					Boltaks
			5×10^{-5}					900	Struthers
			1.5×10^{-5}					1200	Sharma
				0.49 0.45			elec. & opt. elec.		Plotnikov et al., Irmler
	Cu-O				0.24		elec. & opt.		Irmler, Voronkova & Iglitsyn
	Fe	6.2×10^{-3}		0.87				1100-1300	Struthers
			4×10^{-6}					1100	
					0.40 VB 0.55 CB		elec. & opt.		Collins & Carlson
	Ga	3.6		3.5				1100-1350	Fuller & Ditzenberger
			6×10^{-13}					1100	
	Ga-Zn			0.065 VB 0.071 0.083			elec. & opt.		Morin et al., Burstein et al. A, Fuller & Morin
	Ge	6×10^5		5.3				1150-1350	Schibli & Milnes
			5×10^{-13}					1200	Sharma
	H	9.4×10^{-3}		0.48				970-1200	Van Wieringen & Warmoltz
			1.8×10^{-4}					1115	
	He	11×10^{-2}		1.3				970-1200	Van Wieringen & Warmoltz
			6×10^{-6}					1200	
	Hg			0.36 0.31	0.33 VB 0.25		opt.		Zibuts et al. A
	In	16.5		4.0				1100-1350	Fuller & Ditzenberger
			10^{-13}					1100	
				0.15-0.16			elec. & opt.		Burstein et al, Holland & Paul, Newman
				0.155			photocond.		Blakemore & Sarver
	K	1.1×10^{-3}		0.88				830-1100	Svob
			2×10^{-7}					1100	
	Li	2.65×10^{-3}		0.66				400-500	Pratt & Friedman
			2×10^{-7}						

ELECTRICAL PROPERTIES

Diffusion Coeff. and Energy Levels

Dopant	D_o	D	E_{act}	E_a	E_d	Meas.	Temp.(°C)	References
		(cm²/sec)			(eV)			
Li					0.03281	opt.		Gilmer et al., Aggarwal et al.
Li-O					0.039	opt.		Gilmer et al, Aggarwal et al.
Mg					0.1068	opt.		Franks & Robertson
					0.2534			
Mn		2×10^{-7}					1200	Carlson A
					0.53	opt.		Carlson A, Holland
Mo					0.33 CB	opt.		Zibuts et al. A
					0.34 VB			
					0.30 VB			
Na	1.65×10^{-3}		0.70				830-1200	Svob
		7×10^{-7}					1100	
					0.0315	elec.	4-300	Doubrava
N	0.87		3.29					Clark et al.
					0.045	elec.		Clark et al., Zorin et al.
Nd					0.25-0.4	elec.		Gibbons et al.
Ni		1.57×10^{-7}					800	Ridgway & Haneman
	$10^{-4}\text{-}10^{-5}$							Ghandi & Thiel
		3×10^{-8}	2.0				1200	Bonzel
				0.21 VB		elec.	150-400	Ghandi & Thiel
				0.35 CB				
O	0.23		2.5					Corbett et al.
		4.2×10^{-10}					1200	
Si-O$_4$					0.16			Mordkovich C
				0.35		elec.		Mordkovich A,B
				0.38 CB				
P	10.5		3.7				950-1230	Fuller & Ditzenberger
		1.3×10^{-12}					1150	
		2.2×10^{-12}	4.3				1200	Uskov et al.
					0.0453	elec. & opt.		Kohn, Picus et al, Bichard & Giles
					0.055			Tetelbaum
Pt				0.36 VB	0.25 CB	elec.	80-465	Carchano & Jund
				0.30 VB				
S	0.92		2.2				1100-1300	Carlson et al.
		10^{-8}					1100	
				0.105-(0.22P)+(0.25T)			50-300	Camphausen et al.
				0.19-(0.55P)+(0.057T)				
				0.36-(1.1P)-(0.61T)				

ELECTRICAL PROPERTIES REFERENCES

Diffusion Coeff. and Energy Levels	Dopant	D_o	D	E_{act}	E_a	E_d	Meas.	Temp.(°C)	
			(cm^2/sec)		(eV)				
	Neutral S				0.1090		opt.		Krag et al.
					0.1872				
	Singly ionized S				0.3083		opt.		Krag et al.
					0.6116				
	Sb	12.9		3.0				1190-1400	Rohan et al.
			2×10^{-13}					1200	Rohan et al, Sladkov et al., Uskov et al.
			10^{-15}				electron irradiated Si	500-800	Gamo et al.
					0.039		elec. & opt.		Morin et al.
					0.0426				Picus et al.
	Se		$<10^{-8}$					1200	Collins & Carlson
	Si	1.8×10^3		4.8				1200-1400	Peart
			5×10^{-13}					1300	Peart, Masters & Fairfield B
		9×10^3		5.0				1100-1300	Masters & Fairfield B
			10^{-15}					1100	
	Sn	32						1050-1200	Yeh et al.
			8×10^{-14}					1100	
	Te			0.13 VB					Lyutovich
					0.14 CB		elec.		Fischler
	Tl	16.5		4.0				1100-1350	Fuller & Ditzenberger
			8×10^{-14}					1100	
				0.26			elec.		Shulman
	Tm				0.29		elec.		Gibbons et al.
	W			0.37	0.34		opt.		Zibuts et al. A & B
				0.30	0.31				
				0.22					
	Zn	10^{-7}-10^{-6}		1.6				1100-1300	Bakhadyrkhanov et al. A
			$\sim 5 \times 10^{-7}$					1200	Fuller & Morin
				0.31 VB			elec. & opt.		Fuller & Morin, Carlson B
				0.55 CB					
	Zn-Au			0.34			photocond.		Glinchuk et al., Kornilov
				0.55					

ELECTRICAL PROPERTIES	SYMBOL	VALUE	UNIT	NOTES	TEMP.($^\circ$K)	REFERENCES
Energy Gap						
Indirect $(\Gamma_{25'}-X_1)$	E_o	1.200		transmission meas.	0	Balslev
or $(\Gamma_{25'}-\Delta_1)$		1.205	eV	calc. from optical meas.	0	Barber
		1.1558		optical meas.	4.2	Macfarlane et al.
		1.1532			77	
		1.1135			291	
		1.079			415	
		1.122		electroabsorption including 0.01 eV exciton value	296	Frova & Handler, Wendland & Chester
		0.6			1420°C	Glassbrenner & Slack
Spin-orbit Splitting	Δ_o	0.044		optical transmission	300	Zwerdling et al.
Energy Gap						
Direct $(\Gamma_{25'}-\Gamma_{15})$	E_g	3.04		calc.	0	A.R. Williams
Direct	E_g	3.07		Faraday rotation meas. at 20 kGauss, 0.45-2.05 eV,	300	Gabriel
Indirect	E_o	1.18		high purity crystal, oriented normal to (111)		
$(\Delta_5-\Delta_1)$	$E_{o'}$	3.4		wavelength modulation	5	Zucca & Shen
		3.4		electroreflectivity	200	Seraphin & Bottka
		3.2		electroreflectivity	300	Cardona et al.
		3.4		optical reflectivity, high purity	300	Lukes & Schmidt, Seraphin
Spin-orbit Splitting	$\Delta_{o'}$	0.025		piezoelectroreflectance	300	Pollack & Cardona
$(\Lambda_3-\Lambda_1)$	E_1	3.45		wavelength modulation	5	Zucca & Shen
$(\Sigma_3-\Sigma_2)$	E_2	4.31		electroreflectivity	300	Cardona et al. C
		4.44		wavelength modulation	5	Zucca & Shen
$(\Lambda_3-\Lambda_3)$	E_1'	5.50		wavelength modulation	5	Zucca & Shen
(X_4-X_1)		4.2		electroreflectance	300	Seraphin
Valence Band Width		16.7		absorption at 70-200Å		Tomboulian & Bedo
Energy Gap		1.26		optical meas., amorphous films	300	Brodsky et al.
Temperature Coeff.	dE_o/dT	-2.8	10^{-4}eV/$^\circ$K	optical absorption	4.2-4.5	Macfarlane et al.
	$dE_{o'}/dT$	-2.5		optical absorption	100-300	Lukes & Schmidt
	dE_1/dT	-2.2			80-300	Zucca & Shen
	dE_2/dT	-3.0			100-300	Lukes & Schmidt
Pressure Coeff.	dE_o/dP	-8	10^{-6}eV/kg cm^{-2}	<u>Stress</u> (100) diode current		Bulthuis
		-4.5		(110) meas.		
		-5		(111) P= 10^4 kg/cm^2		
		-9.3		(100) opt. absorption	80	Balslev
		-4.6		(110) P=10^4-10^5 kg/cm^2	80	
		-3.0		(111)	80	
		-9.2		(001)	295	

ELECTRICAL PROPERTIES	SYMBOL	VALUE	UNIT	NOTES	TEMP.($°$K)	REFERENCES
Pressure Coeff.	dE_1/dP	+5.1	10^{-6}eV/kg cm^{-2}		300	Zallen & Paul
	dE_2/dP	+2.9				
	dE/dP	$\sim$+0.2		amorphous layers 0.5µ thick	300	Fuhs & Stuke
Deformation Potential						
Indirect Gap	Ξ_u	9.0	eV	cyclotron resonance	4	Murase et al.
	Ξ_d	-6.0				

ELECTRICAL PROPERTIES

ELECTRICAL PROPERTIES	SYMBOL	VALUE	UNIT	NOTES	TEMP.($^\circ$K)	REFERENCES
Nernst Coeff.		55	10^{-10} Gauss $^\circ$K/V		500	Mette et al.
Ettingshausen Coeff.		17	10^{-7} Gauss Amp/cm $^\circ$K		650	Mette et al.
Nernst-Ettingshausen Coeff.		p- n-				
Transverse		12 -5			125	Akimova A,B
Longitudinal		16 -3.5				
Magnetic Susceptibility		-1.11	10^{-7} cgs		293	Busch & Kern
		1.15		$n_n = 10^{13}$	2-50	Sasaki & Kinoshita
		1.7		$n_n = 10^{20}$, P-doped	2-50	Sasaki & Kinoshita
g-Factor		2.0058			100-400	Wada et al.
		1.99893			4.2	Kodera
Superconducting Transition Temp.		6.7	$^\circ$K	P= 120 kbar		Wittig
Pressure Coeff.		10^{-5}	$^\circ$K/bar			
Electronic Specific Heat		5.54	10^5 Joules/mole $^\circ$K^2		1.2-4.2	Keesom & Seidel

OPTICAL PROPERTIES

OPTICAL PROPERTIES	VALUE	WAVELENGTH(μ)	UNIT	NOTES	TEMP.($^\circ$K)	REFERENCES
Absorption Coeff.	1	10	cm^{-1}		300	Johnson
	2	0.3	10^{-6} cm^{-1}		300	Philipp & Taft
Transmission	55	1.1-6.5	%		300	Salzberg & Villa
Reflectivity	30	2-30				
Refractive Index n	3.4975	1.3570			300	Salzberg & Villa
	3.4320	3.0				
	3.4223	5.0				
	3.4179	10.0				
	3.7	2		amorphous film	300	Brodsky et al.
Dispersion $(1/n-1)(dn/d\lambda)$	0.043	1.2			300	Billard & Hily
	0.013	4				
Temp. Coeff. $(1/n)(dn/dT)$	3.9	3	$10^{-5}/^\circ$K		77-400	Cardona A
Pressure Coeff. $(1/n)(dn/dP)$	-3±2	3	10^{-7}cm^2/kg		300	Cardona A
Piezobirefringence (100) (111)						
$(\varepsilon_1)_\parallel - (\varepsilon_1)_\perp / X$ 1.95 1.05	1		10^{-11}cm^2/dyne		300	Higginbotham et al.
1.75 0.9	2					
Spectral Emissivity	0.7				873	Sato
	0.65				1100	Brekhovsikh
Elasto-optic Constants p_{11}	0.081	3.39			300	Pedinoff & Seguin
p_{12}	0.124					
p_{44}	0.075					
$q_{11}-q_{12}$	2.8	1.4	10^{-6}/kg cm^{-2}		300	Nikitenko & Martynenko
q_{44}	1.75					

AGGARWAL, R.L. et al. Excitation Spectra of Lithium Donors in Silicon and Germanium. PHYS. REV., v. 138, no. 3A, May 1965. p. A882-A893.

AKIMCHENKO, I.P. and V.A. VDOVENKOV. Strain-Energy Constants for Silicon, Determined from Uniaxial Compression Effects on the Fundamental Absorption Bands in the Indirect Transition Region. SOVIET PHYS. SOLID STATE, v. 11, no. 3, Sept. 1969. p. 528-533.

AKIMOVA, Kh. A. Longitudinal and Transverse Nernst-Ettingshausen Effects in n-Type Silicon. SOVIET PHYS. SEMICONDUCTORS, v. 1, no. 12, June 1968. p. 1539-1542.   [A]

AKIMOVA, Kh. A. Thermomagnetic Effects in p-Type Silicon. SOVIET PHYS. SEMICONDUCTORS, v. 2, no. 4, Oct. 1968, p. 459-461. [B]

BACHMANN, R. et al. Temperature Dependence of the Silicon Work Function (In Ger.). HELVETICA PHYS. ACTA., v. 39, no. 7, Nov. 1966. p. 593-594.

BACHMANN, R. Temperature Dependence of the Work Function of Silicon (In Ger.). PHYS. KONDENSIERTE MATERIE, v. 8, no. 1, 1968. p. 31-57.

BADALOV, A.Z. and V.B. SHURMAN. Diffusion of Gold in n-Type Silicon. SOVIET PHYS. SEMICONDUCTORS, v. 3, no. 9, Mar. 1970. p. 1137-1139.

BAIDOV, V.V. and M.B. GITIS. Velocity of Sound in and Compressibility of Molten Germanium and Silicon. SOVIET PHYS. SEMICONDUCTORS, v. 4, no. 5, Nov. 1970. p. 825-826.

BAKHADYRKHANOV, M.K. et al. Diffusion Solubility and Electrical Properties of Zinc in Silicon. SOVIET PHYS. SEMICONDUCTORS, v. 4, no. 5, Nov. 1970. p. 739-743. [A]

BAKHADYRKHANOV, M.K. et al. Diffusion, Solubility and Electrical Properties of Cobalt in Silicon. SOVIET PHYS. SOLID STATE, v. 12, no. 1, July 1970. p. 144-149. [B]

BAKHADYRKHANOV, M.K. et al. Optical Absorption of Silicon Containing Cobalt and Zinc Impurities. SOVIET PHYS. SOLID STATE, v. 11, no. 12, June 1970. p. 3076-3077. [C]

BALSLEV, I. Influence of Uniaxial Stress on the Indirect Absorption Edge in Silicon and Germanium. PHYS. REV., v. 143, no. 2, Mar. 1966. p. 636-647.

BARBER, H.D. Effective Mass and Intrinsic Concentration in Silicon. SOLID STATE ELECTRONICS, v. 10, no. 11, Nov. 1967. p. 1039-1051.

BEATTIE, A.G. and J.E. SCHIRBER. Experimental Determination of the Low Temperature Grueneisen Parameter of Silicon from Pressure Derivatives of Elastic Constants. PHYS. REV., B, ser. 3, v. 1, no. 4, Feb. 1970. p. 1548-1551.

BEILIN, V.M. et al. Elastic Constants of Heavily Doped n-Type Silicon and p-Type Germanium. SOVIET PHYS. SOLID STATE, v. 12, no. 3, Sept. 1970. p. 531-535.

BICHARD, J.W. and J.C. GILES. Optical Absorption Spectra of Arsenic and Phosphorous in Silicon. CANADIAN J. OF PHYS., v. 40, no. 10, p. 1480-1489. Oct. 1962.

BILLARD, P. and C. HILY. Weak Field Optical Systems for Use in the Infrared (In Fr.). ACTA ELECTRONICA, v. 11, no. 3, July 1968. p. 237-325.

BLAKEMORE, J.S. and C.E. SARVER. Photoconductivity Associated with Indium Acceptors in Silicon. PHYS. REV., v. 173, no. 3, Sept. 1968. p. 767-774.

BOLTAKS, B.I. Diffusion in Semiconductors. N.Y. Academic Press. 1963.

BOLTAKS, B.I. et al. The Effect of Gold on the Electrical Properties of Silicon. SOVIET PHYS. SOLID STATE, v. 2, no. 2, Aug. 1960. p. 167-175.

BOLTAKS, B.I. and S.Y. HSUEH. Diffusion, Solubility and the Effect of Silver Impurities on Electrical Properties of Silicon. SOVIET PHYS. SOLID STATE, v. 2, no. 11, May 1961. p. 2382-2388.

BONZEL, H.P. Diffusion of Nickel in Silicon. PHYS. STATUS SOLIDI, v. 20, no. 2, Apr. 1967, p. 493-504.

BREKHOVSKIKH, V.F. Experimental Determination of the Emissive Power of Germanium and Silicon in the Temperature Range 700-1200°K. PROGRESS IN HEAT TRANSFER. Ed. by P.K. KONAKOV. N.Y. Consultants Bureau, 1966. p. 145-150.

BRIDGMAN, P.W. The Resistance of 72 Elements, Alloys and Compounds to 100,000 kg/cm^2. AMERICAN ACAD. OF ARTS AND SCI., PROC., v. 81, 1952. p. 165-251.

BRODSKY, M.H. et al. Structural, Optical and Electrical Properties of Amorphous Silicon Films. PHYS. REV., B, ser. 3, v. 1, no. 6, Mar. 1970. p. 2632-2641.

BULLIS, W.M. et al. Temperature Coefficient of Resistivity of Silicon and Germanium Near Room Temperature. SOLID STATE ELECTRONICS, v. 11, no. 7, July 1968. p. 639-646.

BULTHUIS, K. Effect of High Uniaxial Pressure Along the Main Crystallographic Axes for Silicon and Germanium. PHILIPS RES. REPTS., v. 23, no. 1, Feb. 1968. p. 25-47.

BURSTEIN, E. et al. Optical Investigations of Impurity Levels in Silicon. J. OF PHYS. CHEM., v. 57, no. 8, Nov. 1953. p. 849-852. [A]

BURSTEIN, E. et al. Absorption Spectra of Impurities in Silicon. I. Group III Acceptors. PHYS. AND CHEM. OF SOLIDS, v. 1, no. 1/2, Sept./Oct. 1956. p. 65-74. [B]

BURTON, L.C. and A.H. MADJID. Coulomb Screening in Intrinsic Medium-Gap Semiconductors and the Electrical Conductivity of Silicon at Elevated Temperatures. PHYS. REV., v. 185, no. 3, Sept. 1969. p. 1127-1132.

BUSCH, G.A. and R. KERN. The Magnetic Properties of the III-V Compounds (In Ger.). HELVETICA PHYSICA ACTA, v. 32, no. 1, Mar. 1959. p. 24-57.

CAMPHAUSEN, D.L. et al. Influence of Hydrostatic Pressure and Temperature on the Deep Donor Levels of Sulfur in Silicon. PHYS. REV., B, v. 2, no. 6, Sept. 1970. p. 1899-1917.

CARCHANO, H. and C. JUND. Electrical Properties of Silicon Doped with Platinum. SOLID STATE ELECTRONICS, v. 13, no. 1, Jan. 1970. p. 83-89.

CARDONA, M. et al. Dielectric Constant of Germanium and Silicon as a Function of Volume. PHYS. AND CHEM. OF SOLIDS, v. 8, Jan. 1959. p. 204-206. [A]

CARDONA, M. et al. Dielectric Constant Measurements in Germanium and Silicon at Radio Frequencies, as a Function of Temperature and Pressure. In SOLID STATE PHYS. IN ELECTRONICS AND TELECOMM., PROC., Internat. Conf., Brussels, 1958. Ed. DESIRANT, M. and J.L. MICHIELS. N.Y. Academic Press, 1960., v. 1, pt. 1, p. 206-214. [B]

CARDONA, M. et al. Electroreflectance at a Semiconductor-Electrolyte Interface. PHYS. REV., v. 154, no. 3, Feb. 1967. p. 696-720. [C]

CARLSON, R.O. Properties of Silicon Doped with Manganese. PHYS. REV., v. 104, no. 4, Nov. 1956. p. 937-941. [A]

CARLSON, R.O. Double-Acceptor Behavior of Zinc in Silicon. PHYS. REV., v. 108, no. 6, Dec. 1957. p. 1390-1393. [B]

CARLSON, R.O. et al. Sulfur in Silicon. PHYS. AND CHEM. OF SOLIDS, v. 8, Jan. 1959. p. 81-83.

CARR, R.H. et al. Thermal Expansion of Germanium and Silicon at Low Temperatures. PHIL. MAG., v. 12, 8th Series, July 1965. p. 157-163.

CECCHI, R. et al. Electron and Hole Drift Velocity in Silicon at Very Low Temperatures. SOLID STATE COMM., v. 6, no. 10, Oct. 1968. p. 727-728.

CHITTICK, R.C. et al. The Preparation and Properties of Amorphous Silicon. ELECTROCHEM. SOC., J., v. 116, no. 1, Jan. 1969. p. 77-81.

CLARK, A.H. et al. Nitrogen Donor Level in Silicon. AMER. PHYS. SOC., BULL., ser. 11, v. 13, no. 3, Mar. 1968. p. 376.

COLLINS, C.B. et al. Properties of Gold-Doped Silicon. PHYS. REV., v. 105, no. 4, Feb. 1957. p. 1168-1173.

COLLINS, C.B. and R.O. CARLSON. Properties of Silicon Doped with Iron and Copper. PHYS. REV., v. 108, no. 6, Dec. 1957. p. 1409-1414.

CORBETT, J.W. et al. The Configuration and Diffusion of Isolated Oxygen in Silicon and Germanium. PHYS. AND CHEM. OF SOLIDS, v. 25, no. 8, Aug. 1964. p. 873-879.

COSTATO, M. and L. REGGIANI. Electron Drift Velocity and Related Phenomena in Silicon. PHYS. STATUS SOLIDI, v. 42, no. 2, Dec. 1970. p. 591-602.

DONNAY, J.D.H. (Ed.) Crystal Data. Determinative Tables. 2nd Ed. American Crystallography Assoc., 1963.

DOUBRAVA, P. Behaviour of Sodium Donors in Silicon. PHYS. STATUS SOLIDI, v. 34, no. 1, July 1969. p. K9-K11.

DRESSELHAUS, G. et al. Cyclotron Resonance of Electrons and Holes in Silicon and Germanium Crystals. PHYS. REV., v. 98, no. 2, Apr. 1955. p. 368-384.

DUMIN, D.J. and E.C. ROSS. Temperature Dependence of the Hall Mobility and Carrier Concentration in Silicon-on-Sapphire Films. J. APPL. PHYS., v. 41, no. 7, June 1970. p. 3139-3143.

DUNLAP, W.C. and R.L. WATTERS. Direct Measurement of the Dielectric Constants of Silicon and Germanium. PHYS. REV., v. 92, no. 6, Dec. 1953. p. 1396-1397.

STANFORD UNIV., CALIF. Photoemission Studies of the Electronic Band Structures of Gallium Arsenide, Gallium Phosphide and Silicon. TR no. 5221-1 by EDEN, R.C. Contract DA-44-009-AMC-1474 (T), SD-87. May 1967. 303 p. AD 675-474.

EZZ-EL-ARAB, M. et al. Variation in the Speed of Ultrasonic Propagation in Single Crystal Silicon between 25 and 830°C. SOLID STATE COMM., v. 6, no. 6, June 1968. p. 387-390.

FINKE, G. and G. LAUTZ. Low Temperature Behaviour in the Magnetic Resistivity of Silicon Single Crystals (In Ger.). ZEITSCHRIFT FUER NATURFORSCHUNG, v. 11a, no. 12, Dec. 1956. p. 1011-1015.

FISCHLER, S. Vapor Growth and Doping of Silicon Crystals with Tellurium as Carrier. In METALL. OF ADVANCED ELECTRONIC MATERIALS, PROC., Philadelphis, Aug. 1962. N.Y. Interscience, 1963. p. 273-281.

FLUBACHER, P. et al. The Heat Capacity of Pure Silicon and Germanium and Properties of Their Vibrational Frequency Spectra. PHIL. MAG., v. 4, no. 39, Mar. 1959. p. 273-294.

FRANKS, R.K. and J.B. ROBERTSON. Magnesium as a Donor Impurity in Silicon. SOLID STATE COMM., v. 5, no. 6, June 1967. p. 479-481.

FROVA, A. and P. HANDLER. Direct Observation of Phonons in Silicon by Electric-Field Modulated Optical Absorption. PHYS. REV. LETTERS, v. 14, no. 6, Feb. 1965. p. 178-180.

FUHS, W. and J. STUKE. Elastoresistivity of Amorphous Semiconductors. MAT. RES. BULL., v. 5, no. 8, Aug. 1970. p. 611-620.

FULKERSON, W. et al. Thermal Conductivity, Electrical Resistivity and Seebeck Coefficient of Silicon from 100 to 1300°K. PHYS. REV., v. 167, no. 3, Mar. 1968. p. 765-782.

FULLER, C.S. and J.A. DITZENBERGER. Diffusion of Donor and Acceptor Elements in Silicon. J. APPL. PHYS., v. 27, no. 5, May 1956. p. 544-553.

FULLER, C.S. and F.H. MORIN. Diffusion and Electrical Behavior of Zinc in Silicon. PHYS. REV., v. 105, no. 2, Jan. 1957. p. 379-384.

GABRIEL, C.J. Nonresonant Interband Faraday Rotation in Silicon. PHYS. REV., B, v. 2, no. 6, Sept. 1970. p. 1812-1817.

GERHARDT, U. Reflectivity of Germanium and Silicon Single Crystals under Elastic Deformation. (In Ger.) PHYS. STATUS SOLIDI, v. 11, no. 2, 1965. p. 801-818.

GAMO, K. et al. Enhanced Diffusion of High-Temperature Ion-Implanted Antimony into Silicon. APPL. PHYS. LETTERS, v. 17, no. 9, Nov. 1970. p. 391-393.

GHANDI, S.K. et al. Impact Ionization in Cobalt-Doped Silicon. IEEE PROC., v. 53, no. 6, June 1965. p. 635.

GHANDI, S.K. and F.L. THIEL. The Properties of Nickel in Silicon. IEEE PROC., v. 57, no. 9, Sept. 1969. p. 1484-1489.

GIBBONS, D.F. Thermal Expansion of Some Crystals with Diamond Structure. PHYS. REV., v. 112, no. 1, Oct. 1958. p. 136-140.

GIBBONS, J.F. et al. The Doping of Semiconductors by Ion Bombardment. NUCL. INSTRUMENTS AND METHODS., v. 38, 1965. p. 165-168.

GILMER, T.E. Jr. et al. An Optical Study of Lithium and Lithium-Oxygen Complexes as Donor Impurities in Silicon. J. PHYS. CHEM. OF SOLIDS, v. 26, no. 8, Aug. 1965. p. 1195-1204.

GLASSBRENNER, C.J. and G.A. SLACK. Thermal Conductivity of Silicon and Germanium from 3°K to the Melting Point. PHYS. REV., v. 134, no. 4A, May 1964. p. A1058-A1069.

GLAZOV, V.M. et al. Thermal Expansion of Substances Having a Diamondlike Structure and the Volume Changes Accompanying Their Melting. RUSSIAN J. OF PHYS. CHEM., v. 43, no. 2, Feb. 1969. p. 201-205.

GLINCHUK, K.D. et al. Photoconductivity of Silicon with Gold and Zinc Impurities (In Russ.). UKRAINSKII FIZ. ZH., v. 11, no. 12, Dec. 1966. p. 1324-1331.

GOBELI, G.W. and F.G. ALLEN. Photoelectric Properties of Cleaved Gallium Arsenide, Gallium Antimonide, Indium Arsenide and Indium Antimonide Surfaces, Comparison with Silicon and Germanium. PHYS. REV., v. 137, no. 1A, Jan. 1965. p. A245-A254.

GMELINS HANDBUCH DER ANORGANISCHEN CHEMIE: achte völlig neu bearbeitete Auflage. Silicium. Teil B. Weinheim, Verlag Chemie, GmbH, 1959.

GORYUNOVA, N.A. The Chemistry of Diamond-like Semiconductors. Ed. J.C. ANDERSON, The M.I.T. Press, Mass. Inst. of Tech., Cambridge, Mass. 1963, 236 p.

HENSEL, J.C. Cyclotron Resonance in Uniaxially Stressed Silicon. II. Nature of the Covalent Bond. PHYS. REV., v. 138, no. 1A, Apr. 1965. p. A225-A238.

HIGGINBOTHAM, C.W. et al. Intrinsic Piezobirefringence of Germanium, Silicon and Gallium Arsenide. PHYS. REV., v. 184, no. 3, Aug. 1969. p. 821-829.

HOFFMANN, A. et al. Measurement of the Hall Effect and Conductivity of Super-Pure Silicon. J. PHYS. AND CHEM. OF SOLIDS, v. 11, no. 3/4, Oct. 1959. p. 284-287.

HOLLAND, M.G. and W. PAUL. Effect of Pressure on the Energy Levels of Impurities in Semiconductors. I. Arsenic, Indium and Aluminum in Silicon. PHYS. REV., v. 128, no. 1, Oct. 1962. p. 30-38.

HARVARD UNIV., GORDON McKAY LAB. OF APPL. SCI. High Pressure Effects on Impurity Energy Levels in Semiconductors. by HOLLAND, M.G. TR no. HP-4, Contract Nonr-186610. Dec. 1958. AD 213 822.

HONIG, R.E. Vapour Pressure Data for the More Common Elements. RCA REV., v. 18, no. 2, June 1957. p. 195-204.

HURST, W.S. and D.R. FRANKL. Thermal Conductivity of Silicon in the Boundary Scattering Regime. PHYS. REV., v. 186, no. 3, Oct. 1969. p. 801-810.

IBACH, H. Thermal Expansion of Silicon and Zinc Oxide. I. PHYS. STATUS SOLIDI, v. 31, no. 2, Feb. 1969. p. 625-634.

IRMLER, H. Deep Energy Levels in Silicon (In Ger.). ZEITSCHRIFT FUER NATURFORSCHUNG, v. 13a, no. 7, July 1958. p. 557-559.

JAEGER, H. and W. KOSAK. The Metal-Semiconductor-Contact Barriers of Metals from the First and Eight Side Groups, on Silicon and Germanium (In Ger.). SOLID STATE ELECTRONICS, v. 12, no. 7, July 1969. p. 511-518.

JAYARAMAN, A. et al. Melting and Polymorphism at High Pressures in Some Group IV Elements and III-V Compounds with Diamond/Zincblende Structure. PHYS. REV., v. 130, no. 2, Apr. 1963. p. 540-547.

JOHNSON, F.A. Lattice Absorption Bands in Silicon. PHYS. SOC., PROC., v. 73, Feb. 1959, p. 265-272.

JØRGENSEN, M.H. et al. Hot Electron Effects in Silicon and Germanium. PHYSICS OF SEMICONDUCTORS, PROC., 7th Intern. Congress, Paris 1964. N.Y. Academic Press, v. 1, 1964. p. 457-466.

KEESOM, P.H. and G. SEIDEL. Specific Heat of Germanium and Silicon at Low Temperatures. PHYS. REV., v. 113, no. 1, Jan. 1959. p. 33-39.

KIENDL, H. Absolute Precision Determination of Lattice Constants in Single Crystal Silicon with Electron Interference (In Ger.). ZEITSCHRIFT FUER NATURFORSCHUNG, v. 22A, no. 1, Jan. 1967. p. 79-91.

KODERA, H. Effect of Doping on the Electron Resonance in Phosphorous-Doped Silicon. PHYS. OF SEMICONDUCTORS, PROC., Intern. Conf., Kyoto, 1966. Pub. by the Phys. Soc. of Japan, Tokyo, 1966. p. 578-581.

KOHN, W. Shallow Impurity States in Silicon and Germanium. In SOLID STATE PHYSICS, Ed. SEITZ, F. and D. TURNBULL. N.Y. Academic Press, 1957. v. 5, p. 257-320.

KORNILOV, B.V. Absorption in Silicon Doped with Zinc. SOVIET PHYS. SOLID STATE, v. 5, no. 11, May 1964, p. 2420-2424.

KRAG, W.E. et al. Sulfur Donors in Silicon; Infrared Transitions and Effects of Calibrated Uniaxial Stress. PHYS. OF SEMICONDUCTORS, PROC., Intern. Conf., Kyoto, 1966. Pub. by the Phys. Soc. of Japan, Tokyo, 1966. p. 230-233.

LANDOLT-BÖRNSTEIN. Numerical Data and Functional Relationships in Science and Technology, New Series. Ed. K.-H. Hellwege. Group III, Crystal and Solid State Physics, v. 2, pt. 2, N.Y. Springer Verlag. 1969.

LANDWEHR, G. Application of Piezoresistive Effect in Silicon (In Ger.). ZEITSCHRIFT FUER ANGEWANDTE PHYSIK, v. 14, no. 7, July 1962. p. 430-434.

LEADON, R. and J.A. NABER. Recombination Lifetimes in High-Purity Silicon at Low Temperatures. J. APPL. PHYS., v. 40, no. 6, May 1969. p. 2633-2638.

LOGAN, R.A. and A.J. PETERS. Impurity Effects upon Mobility in Silicon. J. APPL. PHYS., v. 31, no. 1, Jan. 1960. p. 122-124.

LOUDON, R. and F.A. JOHNSON. Critical Point Analysis of the Phonon Spectra of Diamond, Silicon and Germanium. In Intern. Conf. on Semiconductors Phys., Proc., 7th, Paris, 1964, v. 1. N.Y. Academic Press, 1964. p. 1037-1049.

LUDWIG, G.E. and R.L. WATTERS. Drift and Conductivity Mobility in Silicon. PHYS. REV., v. 101, no. 6, Mar. 1956, p. 1699-1701.

LUKES, F. and E. SCHMIDT. Reflectivity of Pure and Heavily Doped Silicon in the Energy Range 0.1 to 6 eV. In INTERN. CONF. ON SEMICONDUCTOR PHYS., PROC., 7th, Paris, 1964. v. 1. N.Y. Academic Press, 1964. p. 197-202.

LYUTOVICH, A.S. et al. Some Properties of Tellurium Doped Silicon. SOVIET PHYS. SEMICONDUCTORS, v. 2, no. 6, Dec. 1968. p. 728-729.

McSKIMIN, H.J. and P. ANDREATCH, Jr. Elastic Moduli of Silicon vs. Hydrostatic Pressure at 25.0°C and -195.8°C. J. APPL. PHYS. v. 35, no. 7, July 1964. p. 2161-2165. [A]

McSKIMIN, H.J. and P. ANDREATCH, Jr. Measurement of Third Order Moduli of Silicon and Germanium. J. APPL. PHYS., v. 35, no. 11, Nov. 1964. p. 3312-3319. [B]

MacFARLANE, G.G. et al. Fine Structure in the Absorption Edge Spectrum of Silicon. PHYS. REV., v. 111, no. 5, Sept. 1958. p. 1245-1254.

MAISSEL, L. Thermal Expansion of Silicon. J. APPL. PHYS., v. 31, no. 1, Jan. 1960. p. 2111.

MASTERS, B.J. and J.M. FAIRFIELD. Arsenic Isoconcentration Diffusion Studies in Silicon. J. APPL. PHYS., v. 40, no. 6, May 1969. p. 2390   [A]

MASTERS, B.J. and J.M. FAIRFIELD. Silicon Self Diffusion. APPLIED PHYS. LETTERS, v. 8, no. 11, June 1966. p. 280-281. [B]

NATIONAL BUREAU OF STANDARDS. ELECTRONIC INSTRUMENTATION DIV., GAITERSBURG, MD. Measurement and Interpretations of Carrier Lifetime in Silicon and Germanium. by MATTIS, R.L. et al. FR. Dec. 1, 1966-Jan. 1, 1968. Contract no. F 33 615-M-5007. July 1968. 91 p.

METALS HANDBOOK 8th Ed., v. 1. Properties and Selection of Metals. p. 1222. Metals Park, Novelty, Ohio. Ed. T. Lyman. Amer. Soc. for Metals, 1961.

MESSIER, J. and J.M. FLORES. Temperature Dependence of Hall Mobility and the Ratio of the Hall Mobility to the Drift Mobility for Silicon. J. OF PHYS. AND CHEM. OF SOLIDS, v. 24, no. 12, Dec. 1963. p. 1539-1542.

METTE, H. et al. Nernst and Ettingshausen Effects in Silicon between 300 and 800°K. PHYS. REV., v. 117, no. 6, Mar. 1960. p. 1491-1493.

METZGER, H. and F.R. KESSLER. The Debye-Sears Effect for Determination of the Silicon Elastic Constants (In Ger.). ZEITSCHRIFT FUER NATURFORSCHUNG, v. 25a, no. 6, June 1970. p. 904-908.

MILLER, R.C. and A. SAVAGE. Diffusion of Aluminum in Single Crystal Silicon. J. APPL. PHYS., v. 27, no. 12, Dec. 1956. p. 1430-1432.

MOORE, J.S. et al. Energy Levels in Cobalt Compensated Silicon. J. APPL. PHYS., v. 41, no. 13, Dec. 1970. p. 5282-5285.

MORDKOVICH, V.N. The Influence of Oxygen on the Conductivity of Silicon. SOVIET PHYS. SOLID STATE, v. 6, no. 3, Sept. 1964. p. 654-657. [A]

MORDKOVICH, V.N. The Effect of Oxygen on the Electrical Properties of n-Type Silicon. SOVIET PHYS. SOLID STATE, v. 4, no. 12, p. 2662-2664, June 1963. [B]

MORDKOVICH, V.N. Effect of Oxygen on Recombination in Silicon. SOVIET PHYS. SOLID STATE, v. 6, no. 7, Jan. 1965. p. 1716-1717. [C]

MORIN, F.J. and J.P. MAITA. Electrical Properties of Silicon Containing Arsenic and Boron. PHYS. REV., v. 96, no. 1, Oct. 1954. p. 28-35.

MORIN, F.J. et al. Impurity Levels in Silicon. PHYS. REV., v. 96, no. 3, Nov. 1954. p. 833.

MURASE, K. et al. Determination of Deformation Potential Constants from the Electron Cyclotron Resonance in Germanium and Silicon. PHYS. SOC. OF JAPAN, J., v. 29, no. 5, Nov. 1970. p. 1248-1257.

NATHAN, M.I. and W. PAUL. Effect of Pressure on the Energy Levels of Impurities in Semiconductors. II. Gold in Silicon. PHYS. REV., v. 128, no. 1, Oct. 1962. p. 38-42.

NEWMAN, R. Optical Properties of Indium-Doped Silicon. PHYS. REV., v. 99, no. 2, July 1955. p. 465-467.

NEWMAN, R.C. and J. WAKEFIELD. The Diffusivity of Carbon in Silicon. J. OF THE PHYS. AND CHEM. OF SOLIDS, v. 19, no. 3/4, 1961. p. 230-234.

NIKITENKO, V.I. and G.P. MARTYNENKO. Certain Photoelastic Properties of Gallium Arsenide and Silicon. SOVIET PHYS. SOLID STATE, v. 7, no. 2, Aug. 1965. p. 494-496.

NOACK, J. Measurement of Minority Carrier Lifetime in Silicon of Low Dislocation Density. PHYS. STATUS SOLIDI, v. 32, no. 1, Mar. 1969. p. K17-K19.

OHYAMA, T. et al. Observation of Cyclotron Resonance at the Split-Off Valence Band in Silicon. PHYS. LETTERS, v. 33A, no. 1, Sept. 1970. p. 55-57.

PEART, Q.F. Self Diffusion in Intrinsic Silicon. PHYS. STATUS SOLIDI, v. 15, no. 1, 1966. p. K119-K122.

PEDINOFF, M.E. and H.A. SEGUIN. Direct Measurements of Infrared Photo-Elastic Constants of Silicon. IEEE J. OF QUANTUM ELECTRONICS, v. QE-3, no. 1, Jan. 1967. p. 31-32.

PENCHINA, C.M. et al. Energy Levels and Negative Photoconductivity in Cobalt-Doped Silicon. PHYS. REV., v. 143, no. 2, Mar. 1966. p. 634-636.

PETERSON, C.W. et al. Photoemission from Amorphous Silicon. PHYS. REV. LETTERS, v. 25, no. 13, Sept. 1970. p. 861-864.

PHILIPP, H.R. and E.A. TAFT. Optical Constants of Silicon in the Region 1 to 10 eV. PHYS. REV., v. 120, no. 1, Oct. 1960. p. 37-38.

PICUS, G. et al. Absorption Spectra of Impurities in Silicon. II. Group V Donors. J. OF PHYS. AND CHEM. OF SOLIDS, v. 1, no. 1/2, 1956. p. 75-81.

PLOTNIKOV, A.F. et al. Remanent Impurity Photoconductivity Spectra of Silicon Single Crystals. SOVIET PHYS. SOLID STATE, v. 4, no. 12, June 1963. p. 2616-2617.

POLLAK, F.H. and M. CARDONA. Piezo-Electroreflectance in Germanium, Gallium Arsenide and Silicon. PHYS. REV., v. 172, no. 3, Aug. 1968. p. 816-837.

PRATT, B. and F. FRIEDMAN. Diffusion of Lithium into Germanium and Silicon. J. APPL. PHYS., v. 37, no. 4, Mar. 1966. p. 1893-1896.

QUARANTA, A.A. et al. Experimental Results on the Drift Velocity of Hot Carriers in Silicon and Associated Anisotropic Effects. SOLID STATE ELECTRONICS, v. 11, no. 7, July 1968. p. 685-696.

RAO, K.V. and A. SMAKULA. Dielectric Anomalies in Silicon Single Crystals. J. APPL. PHYS., v. 37, no. 7, June 1966. p. 2840-2842.

RIDGWAY, J.W.T. and D. HANEMAN. The Diffusion of Iron and Nickel to Silicon Surfaces. PHYS. STATUS SOLIDI, v. 38, no. 1, Mar. 1970. p. K31-K33.

ROHAN, J.J. et al. Diffusion of Radioactive Antimony in Silicon. ELECTROCHEM. SOC., J., v. 106, no. 8, Aug. 1959. p. 705-709.

SALZBERG, C.D. and J.J. VILLA. Infrared Refractive Indices of Silicon, Germanium and Modified Selenium Glass. OPTICAL SOC. OF AMERICA, J., v. 47, no. 3, Mar. 1957. p. 244-246.

SASAKI, W. and J. KINOSHITA. Piezoresistance and Magnetic Susceptibility in Heavily Doped n-Type Silicon. PHYS. SOC. OF JAPAN, J., v. 25, no. 6, Dec. 1968. p. 1622-1629.

SATO, T. Spectral Emissivity of Silicon. JAPAN. J. OF APPL. PHYS., v. 6, no. 3, Mar. 1967. p. 339-347.

SCHIBLI, E. and A.G. MILNES. Deep Impurities in Silicon. MATERIALS SCI. AND ENG., v. 2, no. 4, Nov. 1967. p. 173-180.

SCHLOETTERER, H. Mechanical and Electrical Properties of Epitaxial Silicon Films on Spinel. SOLID STATE ELECTRONICS, v. 11, no. 10, Oct. 1968. p. 947-956.

SERAPHIN, B.O. Optical Field Effect in Silicon. PHYS. REV., v. 140, no. 5A, Nov. 1965. p. A1716-A1725.

SERAPHIN, B.O. and N. BOTTKA. Band Structure Analysis from Electroreflectance Studies. PHYS. REV., v. 145, no. 2, May 1966. p. 628-636.

SHAKLEE, K.L. and R.E. NAHORY. Valley-Orbit Splitting of Free Excitons? The Absorption Edge of Silicon. PHYS. REV. LETTERS, v. 24, no. 17, Apr. 1970. p. 942-945.

SHANKS, H.R. et al. Thermal Conductivity of Silicon from 300-1400°K. PHYS. REV., v. 130, no. 5, June 1963. p. 1743-1748.

SHARMA, B.L. Diffusion in Semiconductors. TRANS. TECH. PUBL. Adolf Ey Str. 1d D3392. Clausthal Zellerfeld, Germany.

SHULMAN, R.G. Tight-Bonding Calculation of Acceptor Energies in Germanium and Silicon. J. OF THE PHYS. AND CHEM. OF SOLIDS, v. 2, no. 2, 1957. p. 115-118.

SLADKOV, I.B. et al. Distribution of Antimony in Homoepitaxial Silicon Films Prepared by the Silane and Chloride Methods. SOVIET PHYS. SEMICONDUCTORS, v. 4, no. 4, Oct. 1970. p. 676-677.

SMITH, C.S. Piezoresistance Effect in Germanium and Silicon. PHYS. REV., v. 94, no. 1, Apr. 1954. p. 42-49.

SPARKS, C.W. and C.A. SWENSON. Thermal Expansion from 2 to 40°K of Germanium, Silicon and Four III-V Compounds.
PHYS. REV., v. 163, no. 2, Nov. 1967. p. 779-790.

STRUTHERS, J.D. Solubility and Diffusivity of Gold, Iron and Copper in Silicon. J. APPL. PHYS., v. 27, no. 12,
Dec. 1956. p. 1560.

SU, J.L. et al. High-Field Drift Velocity of Electrons in p-Type Silicon. SOLID STATE ELECTRONICS, v. 13, no. 7,
July 1970. p. 1115-1117.

SVOB, L. Solubility and Diffusion Coefficient of Sodium and Potassium in Silicon. SOLID STATE ELECTRONICS, v. 10
no. 10, Oct. 1967. p. 991-996.

TAFT, E.A. and R.O. CARLSON. Beryllium as an Acceptor in Silicon. ELECTROCHEM. SOC., J., v. 117, no. 5, May 1970.
p. 711-713.

TASCH, A.F. Jr. and C.T. SAH. Recombination-Generation and Optical Properties of Gold Acceptor in Silicon.
PHYS. REV., B, ser. 3, v. 1, no. 2, Jan. 1970. p. 800-809.

TETELBAUM, D.I. Temperature Dependence of the Carrier Density and Mobility in Silicon Doped with Boron and Phosphorous by the Ion Implantation Method. SOVIET PHYS. SEMICONDUCTORS, v. 1, no. 5, Nov. 1967. p. 593-597.

THIEL, F.L. and S.K. GHANDHI. Electronic Properties of Silicon Doped with Silver. J. APPL. PHYS., v. 41, no. 1,
Jan. 1970. p. 254-263.

TOMBOULIAN, D.H. and D.E. BEDO. Absorption and Emission Spectra of Silicon and Germanium in the Soft X-Ray
Region. PHYS. REV., v. 104, no. 3, Nov. 1956. p. 590-597.

TUFTE, O. N. and E.L. STELZER. Piezoresistive Properties of Heavily Doped n-Type Silicon. PHYS. REV., v. 133,
no. 6A, Mar. 1964. p. A1705-A1716.

USKOV, V.A. et al. Diffusion of Antimony, Phosphorous and Boron in Silicon with Various Surface Concentrations of
the Diffusant. SOVIET PHYS. SOLID STATE, v. 12, no. 5, Nov. 1970. p. 1181-1185.

VAN WIERINGEN, A. and N. WARMOLTZ. On the Permeation of Hydrogen and Helium in Single Crystal Silicon and Germanium
at Elevated Temperatures. PHYSICA, v. 22, 1956. p. 849-865.

VERESHCHAGIN, L.F. et al. Variation of the Electrical Resistance of Some Semiconductors under Pressures up to
300,000 kg/cm^2. SOVIET PHYS. DOKLADY, v. 7, no. 8, Feb. 1963. p. 692-694.

VORONKOVA, G.I. and M.I. IGLITSYN. Recombination at Copper-Oxygen Complexes in Silicon. SOVIET PHYS. SEMICON-
DUCTORS, v. 1, no. 1, July 1967. p. 120-121.

WADA, T. et al. Annealing Effects of Paramagnetic Defects Introduced near Silicon Surface. PHYS.SOC. OF JAPAN,
J., v. 22, no. 4, Apr. 1967. p. 1060-1065.

WEBER, R.E. and W.T. 'PERIA. Work Function and Structural Studies of Alkali-Covered Semiconductors. SURFACE SCI.,
v. 14, no. 1, 1969. p. 13-38.

WENDLAND, P.H. and M. CHESTER. Electric Field Effects on Indirect Optical Transitions in Silicon. PHYS. REV.,
v. 140, no. 4A, Nov. 1965. p. A1384-A1390.

WILLIAMS, E.L. Boron Diffusion in Silicon. ELECTROCHEM. SOC., J. v. 108, Aug. 1961. p. 795-798.

WILLIAMS, A.R. Non-Muffin-Tin Energy Bands for Silicon by the Korringa-Kohn-Rostoker Method. PHYS. REV., B,
ser. 3, v. 1, no. 8, Apr. 1970. p. 3417-3426.

WITTIG, J. Superconductivity of Germanium and Silicon under High Pressure. (In Ger.) ZEITSCHRIFT FUER PHYSIK,
v. 195, no. 2, 1966. p. 215-227.

WOLFF, G.A. et al. Relationship of Hardness, Energy Gap and Melting Point of Diamond Type and Related Structures.
In SEMICONDUCTORS AND PHOSPHORS, PROC., Intern. Colloquium 1956. Garmisch-Partenkirchen. Ed. by SCHON, M.and H.
WELKER. N.Y. Interscience, 1958. p. 463-469.

WORTMAN, J.J. and R.A. EVANS. Young's Modulus, Shear Modulus and Poisson's Ratio in Silicon and Germanium. J.
APPL. PHYS., v. 36, no. 1, Jan. 1965. p. 153-156.

WYCKOFF, R.W.G. Crystal Structures. N.Y. Wiley & Sons, v. 1, p. 114.

YEH, T.H. et al. Diffusion of Tin into Silicon. J. APPL. PHYS., v. 39, no. 9, Aug. 1968. p. 4266-4271.

ZADUMKIN, S.N. Surface Tension and Heat of Sublimation of Silicon, Germanium and Tin. SOVIET PHYS. SOLID STATE,
v. 1, no. 3, Mar. 1959. p. 516-517.

ZALLEN, R. and W. PAUL. Effect of Pressure on Interband Reflectivity Spectra of Germanium and Related Semi-conductors. PHYS. REV., v. 155, no. 3, Mar. 1967. p. 703-711.

ZIBUTS, Yu. A. et al. Some Properties of Silicon Containing Mercury, Tungsten, Molybdenum and Platinum Impurities. SOVIET PHYS. SOLID STATE, v. 5, no. 11, May 1964. p. 2416-2419. [A]

ZIBUTS, Yu. A. et al. Photoelectric Properties of Silicon Doped with Copper, Tungsten and Platinum. SOVIET PHYS. SOLID STATE, v. 8, no. 9, Mar. 1967. p. 2041-2047. [B]

ZORIN, E.I. et al. Donor Properties of Nitrogen in Silicon. SOVIET PHYS. SEMICONDUCTORS, v. 2, no. 1, July 1968. p. 111-113.

ZUCCA, R.R.L. and Y.R. SHEN. Wavelength-Modulation Spectra of Some Semiconductors. PHYS. REV., B, ser. 3, v. 1, no. 6, Mar. 1970. p. 2668-2676.

ZWERDLING, S. et al. Internal Impurity Levels in Semiconductors. Experiments in p-Type Silicon. PHYS. REV. LETTERS, v. 4, no. 4, Feb. 1960. p. 173-176.

ZYUZ, L.N. et al. Photostimulated Diffusion in Silicon. JETP LETTERS, v. 12, no. 5, Sept. 1970. p. 147-149.

SILICON CARBIDE

PHYSICAL PROPERTIES	SYMBOL	VALUE	UNIT	NOTES	TEMP.(°K)	REFERENCES
Formula		SiC				
Molecular Weight		40.1				
Density						
2H		3.214	g/cm^3		300	de Mesquita
6H		3.211				
51R		3.218			300	Thibault A
β		3.210			300	de Mesquita
β		3.166			20°C	Kern et al.
		3.125			1000°C	
		3.075			2000°C	
Name		carborundum				
Polytypes		∿75		in hexagonal symmetry: layers formed principally by a screw dislocation mechanism		Shaffer A, Gmelin, p. 784-807
				doping may also cause polytypes		Vakhner & Tairov A
Color						
Hexagonal (α)		colorless		pure, transparent,		Brown Boveri
		green		N-, P-doped		Res. Ctre.,
		blue, black		Al-doped		Nelson et al.,
		brown, black		B-doped		Gmelin
Cubic (β)		light yellow		high-purity		Nelson et al.
Hardness						
Cubic (β)		9.2-9.3	kg/mm^2	Mohs		Kern et al.
	H_{100}	3100-3475		Knoop		
	H_{100}	2600		$\parallel$(100)		Shaffer C
Hexagonal (α)	H_{100}	2130		$\parallel$-c axis		Shaffer B
		2755		$\perp$-c axis		
Cleavage		(001)		hexagonal		Goryunova, p. 71
Symmetry		cubic, zincblende		β-form		Donnay
Space Group		F$\bar{4}$3m Z-4				Donnay
Lattice Parameters	a_o	4.357	Å	β-form		Ziomek & Pickar
		4.3595		high-purity		Taylor & Jones
	Si-C	1.89				Wyckoff, v. 1, p. 114
Symmetry		hexagonal, wurtzite		α-form; 2H		Donnay
Space Group		P6mc Z-4		α-form; 2H		Shaffer A
		C6mc		α-form; 4H, 6H		
Symmetry		rhombohedral		α-form		Wyckoff
Space Group		R3m				Wyckoff, Shaffer A

SILICON CARBIDE

PHYSICAL PROPERTIES	SYMBOL	VALUE		UNIT	NOTES	TEMP. (°C)	REFERENCES
Lattice Parameters	a_o	c_o			α-form		
2H	3.0763	5.0480		Å			Adamsky & Merz
4H	3.076	10.046					Donnay
6H	3.08065	15.11738			high-purity		Taylor & Jones
8H	3.079	20.147					Donnay
15R	3.073	37.30					
21R	3.073	52.78					
51R	3.073	128.17					Thibault A
	a_o	α	Z				
15R, Hexagonal	12.691	13°54'5"	5		α-form		Donnay
21R, Rhombohedral Cell	17.683	9°58'	7				
Melting Point		2830		°C	decomposes		Scace & Slack
		1400		°C	decomposes in O_2		Kulvarskaya et al.
Specific Heat							
Cubic		0.17		cal/gm °C		0	Kern et al.
		0.22				200	
		0.28				1000	
		0.30				1400-2000	
Hexagonal		0.165				27	Brown
		0.27				700	Shaffer D
		0.35				1550	
Debye Temperature		913		°K		53-73°K	Gmelin, p. 820
		1430		°K		400°K	Taylor & Jones
Thermal Conductivity							
Hexagonal		52		W/cm °K	high-purity 6H	50°K	Slack
		15				150°K	
		4.9				300°K	
		0.410				20	Shaffer D
		0.335				600	
		0.255				800	
		0.213				1000	
Cubic		1			polycrystalline	30	Bosch
	$\perp$ (111)	$\parallel$ (111)					
	0.255	0.226				200	Kern et al.
	0.155	0.155				1000	
	0.121	0.138				1400	
	0.125	0.138				2000	
Sintered		0.1				1000	Lezhenin & Gnesin
		0.2				1650	

PHYSICAL PROPERTIES	SYMBOL	VALUE	UNIT	NOTES	TEMP.(°K)	REFERENCES
Thermal Expansion Coeff.		2.47	$10^{-6}/$ °C	polycrystalline, CVD	20°C	Kern et al.
Cubic,Hexagonal		0.3		single crystals of cubic	100	Taylor & Jones
		1.1		and 6H polytypes	200	
		2.9			300	
		4.0			400	
		4.5			1000	
		5.4		cubic, hexagonal a-axis	1500°C	Taylor & Jones
		4.9		hexagonal c-axis	1500°C	
Elastic Coefficient						
Stiffness						
Cubic	c_{11}	3.523	10^{12}dynes/cm^2	calc.	300	Tolpygo
	c_{12}	1.404				
	c_{44}	2.329				
Hexagonal	c_{11}	5.00		6H	300	Arlt & Schodder
	c_{12}	0.92				
	c_{44}	1.68				
Compliance						
Hexagonal	s_{11}	2.03	10^{-13}cm^2/dyne	calc.	300	Arlt & Schodder
	s_{12}	-0.421				
	s_{44}	5.95				
Shear Modulus						
Hexagonal		24.41	10^6psi		20°C	Shaffer D
		27.85		high purity hot-pressed powder	20°C	Carnahan
Bulk Modulus						
Hexagonal		14.01	10^6psi		20°C	Shaffer D
Young's Modulus						
Hexagonal		56	10^6psi		20°C	Shaffer D
		51			1200°C	
		65		high purity hot-pressed powder	20°C	Carnahan
Wave Velocity						
Longitudinal		1.326	10^6 cm/sec.	6H type ∥ c-axis, 70 MHz, T=70-300°K	300	Arlt & Schodder
ELECTRICAL PROPERTIES						
Electrical Resistivity						
Hexagonal		40	ohm-cm	high purity, 6H	100	Barrett
		0.8			250	
		1		$n = 10^{16}$ cm^{-3} at 300°K	300	
		10			900	
Cubic		0.7		high purity $n = 10^{16}$ cm^{-3} at 300°K	300	Nelson et al.

SILICON CARBIDE

ELECTRICAL PROPERTIES	SYMBOL	VALUE	UNIT	NOTES	TEMP.($^\circ$K)	REFERENCES
Dielectric Constant						
Hexagonal						
Static	ε_o	9.66		6H-calc.		Patrick & Choyke
	ε_o	10.32				
	ε_o	10.2		pure single crystals at 100 kHz	20,77	Hofman et al.
Optical	ε_∞	6.7		optical meas. 1-25μ	300	Spitzer et al.
Static	ε_o	10.0				
Cubic						
Static	ε_o	9.72		calc.		Patrick & Choyke
Dissipation Factor						
Hexagonal	$\tan \delta$	3×10^{-3}		100 kHz	20	Hofman et al.
		8×10^{-3}				
Mobility						
Cubic						
Electron	μ_n	980	cm^2/V sec.	high-purity single crystal	300	Nelson et al., Silva et al.
Temp. Coeff.	μ_n	$T^{-0.5}$			<200	Rosengreen

ELECTRICAL PROPERTIES	SYMBOL	4H	6H	15R	NOTES	TEMP.($^\circ$K)	REFERENCES
Hexagonal							
Electron	μ_n	500	1000 300 55	1800 400 80	$n_n = 10^{17}$ cm^{-3}, high-purity	100 300 600	Barrett & Campbell
Temp. Coeff.			$T^{-2.4}$			>300	

ELECTRICAL PROPERTIES	SYMBOL	VALUE	UNIT	NOTES	TEMP.($^\circ$K)	REFERENCES
Hole	μ_p	50	cm^2/V sec.	p-type, Al- or B-doped $n_p = 10^{17}\text{-}10^{18}$	300	Van Daal
Effective Mass						
Cubic						
Electron		0.4	m_o	cyclotron resonance meas.	300	Pickar et al.

ELECTRICAL PROPERTIES	SYMBOL	6H	15R	NOTES	TEMP.($^\circ$K)	REFERENCES
Hexagonal						
Electron, transverse	$m_{n\perp}$	0.25	0.28	absorption meas. at 0.4-5μ	300	Ellis & Moss
longitudinal	$m_{n\parallel}$	1.5	0.53	$n_n = 10^{18}\text{-}10^{19}$ cm^{-3}		

ELECTRICAL PROPERTIES	SYMBOL	VALUE	UNIT	NOTES	TEMP.($^\circ$K)	REFERENCES
Hole	m_p	3.5		electrical meas., 6H	300	Van Daal et al.
Density of States	m_{dp}	1.0				
Lifetime						
Electron		<1	μsec.	electrical meas.	300	Ozarow & Hysell
		$<4\times10^{-3}$		electroluminescence meas.	300	Yu et al.
Piezoelectric Coeff.	e_{33}	0.2	C/m_2	6H type	300	Van Daal
	e_{15}	0.08				
Electromechanical Coupling Coeff.	k_{31}	$<4\times10^{-3}$			300	Landolt Börnstein, v. 2, p. 61
	k_{33}	$<7\times10^{-3}$				

ELECTRICAL PROPERTIES		VALUES				NOTES	TEMP.(°C)	REFERENCES
Diffusion and	Dopant	D_o (cm²/sec)	D	E_{act} (eV)	E_a			
	Al	8		+6.1		Boron diffusion prevented	1800-2400	Mokhov et al.
			2×10^{-13}				1950	
			3×10^{-11}				2400	
			5×10^{-10}				2050	Chang
			10^{-12}				1750	
					0.23	luminescence meas. in cubic	77°K	Zanmarchi A
					0.28	luminescence meas.	88°K	Gorban et al. A,
					0.39	on 6H crystals,		Bukke et al.,
					0.49	Al-doped to 10^{-19} cm⁻³		Van Daal et al.
	B		1.6×10^2	5.6				Vodakov et al.
					0.36	transmission meas., 6H, n_p to 10^{18} cm⁻³	300°K	Pichugin & Pikhtin
	Be (fast)	0.3		3.1		6H, diffusion along (001)	1700-2250	Maslakovets et al.
	(slow)	32.0		5.2				
			10^{-8} (fast)				2100	
			10^{-10} (slow)					
					0.60 3×10^{17}	n_p electrical meas.	100-400°K	Maslakovets et al.
					0.42 8×10^{16}			
					0.32	luminescence meas.	280-435°K	Kalnin et al.
	C	3×10^2		6.14		Al-doped	1850-2060°K	Ghoshtagore & Coble
		2×10^{17}		13.11		N-doped	1970-2090	
	Cr	0.228		4.8			1700-1900	Griffiths
			9×10^{-11}				2150	
			7×10^{-12}				2010	
	Ga				0.35	electroluminescence meas.	300°K	Kholuyanov
	N	$4.6\text{--}8.7 \times 10^4$		$7.6\text{--}9.35$		Al-doped crystals	2000-2550	Kroko & Milnes
			7×10^{-12}				2780°K	
			10^{-16}				2270°K	
			5×10^{-9}				2720°K	Slack & Scace
					0.6	high N-doping, N in Si sites, electro-luminescence meas.	300°K	Kholuyanov
					0.17	photoluminescence in 6H	6°K	Hamilton et al. A
					0.20	absorption on 6H	120-300°K	Vakulenko & Govorova
					0.23	thermoluminescence meas.	100-180°K	Gorban et al. B
						absorption on 6H	120-300°K	Vakulenko & Govorova

ELECTRICAL PROPERTIES		VALUES				NOTES	TEMP.(°C)	REFERENCES
Diffusion and Energy Levels	Dopant	D_o (cm^2/sec)	D	E_{act} (eV)	E_a			
	N				0.14 0.16 0.20	absorption and luminescence in 15R	6°K	Patrick et al. A
					0.08 0.04	Hall meas. on 6H Hall meas. on 15R	100-1000°K	Hagen & Kapteyns
	Sc				0.37, 0.42	luminescence meas. on 6H; Sc-doping yields 4H crystals	300°K	Vakhner & Tairov B
	Si	2×10^{-10}					2060	Ghoshtagore & Coble

ELECTRICAL PROPERTIES	SYMBOL	VALUE	UNIT	NOTES	TEMP.(°K)	REFERENCES
Energy Gap						
Cubic, Indirect	E_g	2.60	eV	optical absorption at 0.3-0.8μ, T=77°K, 300°K, 523°K	0	Ziomek & Pickar
		2.39		optical absorption	4.2	Choyke et al. A
Direct		6.0		optical absorption	300	Choyke & Patrick A
Exciton Energy Gap, $(E_{gx}=E_g+E_x)$ Hexagonal						
Indirect, 2H	E_{gx}	3.330		absorption and luminescence meas. on high purity 2H, T=2-8°K	0	Patrick et al. B
4H		3.265 3.264		optical transmission on 4H	4 77	Choyke et al. B
6H		3.024		absorption meas.	4.2	Pikhtin & Yas'kov, Choyke & Patrick B,C
		2.994 2.832		absorption meas. absorption meas.	300 600	Pikhtin & Yas'kov
		2.09		4-particle nitrogen-exciton complex, absorption meas. E c	86	Gorban & Krokhmal
		3.04		E c		
15R		2.986		absorption meas.	4.2	Choyke et al. C
		2.985		absorption meas.	77	Patrick et al.
33R		3.003		absorption meas.	4.2	Choyke et al. C
21R		2.853 2.852		absorption meas. at T=4.2-300°K	4.2 77	Hamilton et al. B
24R		2.728 2.699 2.667		absorption meas. at T=8-476°K	8 300 476	Zanmarchi B
Indirect, X_5-X_1		5.85	eV	reflectivity, $n_p = 10^{14}$	300	Zhumakulov & Smirnova
Binding Energy, E_x-N		16,31,32.5	meV	exciton-nitrogen centers		Choyke & Patrick B

ELECTRICAL PROPERTIES	SYMBOL	VALUE	UNIT	NOTES	TEMP.(°K)	REFERENCES
Energy Gap						
Temperature Coeff.						
Cubic		-2.2	10^{-4} eV/°K	reflectivity meas. at T=100, 290°K		Belle et al.
		-5.8		transmission meas.	295-700	Dalven
Hexagonal, 6H		-3.3		optical meas. at T=6-200°K		Choyke & Patrick
15R		-3.3		absorption and luminescence meas.	6-650	Patrick et al.
Deformation Potential						
Conduction Band		11.5	eV	6H, electrical meas.		Van Daal et al.
Valence Band Width		18.2	eV	cubic and hexagonal forms, ultra-soft x-ray meas.		Zhukova et al.

Phonon Spectra		4.2°K *	300°K **				
Cubic							
Transverse Optical	TO	94.4	93.6	meV	*optical absorption and luminescence meas.		*Choyke et al. A
Transverse Acoustic	TA	46.3	45.9				
Longitudinal Optical	LO	102.8	101.9		+mobility meas.		+Patrick
Longitudinal Acous.	LA	79.4	78.7+		**optical absorption meas. at 2-25μ and Raman spectra of high purity single crystals		**Ziomek & Pickar

Hexagonal	2H	4H	6H	15R	21R	24R	33R		°K
TA_1	53	56	46.3	46.3	46.5	46.0	46.3	absorption and luminescence meas.	4-2H -Patrick et al. B
TA_2	61.5		53.5	51.9	53		52.3		20-4H -Lisitsa et al.
TO_1	91.2	100	94.7	94.6	94.5	94.4	94.7		6-6H -Choyke & Patrick B, Hamilton et al. A, Pikhtin & Yas'kov
TO_2	103.4		95.6	95.7			95.7		6-15R-Patrick et al. A
LA	61.5	75	77	78.2	77.5	77.3	77.5		6-21R-Hamilton et al. B
LO	100.3	120	104.2	103.7	104	104.5	103.7		300-24R-Zanmarchi B
									6-33R-Choyke et al. C

	SYMBOL	VALUE	UNIT	NOTES	TEMP.(°K)	REFERENCES
Work Function	φ					
Hexagonal		4.02	eV	thermionic emission at 1700-1800°K	300	Kan & Kulvarskaya
		4.4-4.6		electron photoemission	300	Dillon et al.
Electron Affinity	ψ					
Cubic		4		electron photoemission	300	Philipp & Taft
Photoelectric Threshold	φ					
Hexagonal, Cubic		>7		electron photoemission	300	Pong & Fujita, Philipp & Taft

Seebeck Coeff.		300-500°K	100°K			
Hexagonal		-1	-6	mV/°K	$n_n = 10^{16}-10^{18}$	Lomakina & Vodakov
		+1.5	+0.5		$n_p = 10^{17}-10^{18}$	
		-0.7			silicon carbide-carbon rods, resistivity= 0.007 ohm cm	300	Breckenridge
		-0.11				1273	

SILICON CARBIDE

ELECTRICAL PROPERTIES	SYMBOL	VALUE	UNIT	NOTES	TEMP. (°K)	REFERENCES
Magnetic Susceptibility						
Hexagonal 6H		-10.618	10^{-6}/g mol	high purity, $\parallel$ c-axis	300	Das
21R		-5.74				
g-Factor		2.003		amorphous film	77	Brodsky & Title
Transverse		2.010		6H	77	Van Ryneveld &
Longitudinal		2.005				Loubser
Spectral Emissivity		0.94		at 9μ	1800	Durand & Houston
		0.8		at 0.6-4μ	2500	Dubrovskii
		0.2		at 0.6μ	2000	

OPTICAL PROPERTIES

Refractive Index	SYMBOL	VALUE	Wavelength(μ)	NOTES	TEMP. (°K)	REFERENCES
Cubic	n	2.48	0.6	n-type single crystals, light yellow	300	Belle
		2.7104	0.467	high purity, defect free	300	Shaffer & Naum
		2.6525	0.589			
		2.6264	0.691			
Cubic, $n_\lambda = 2.55378 + 3.417\mathrm{x}10^{4}/\lambda^{2}$			0.467-0.691		300	Shaffer & Naum
Dispersion		35°36'	0.589			

Hexagonal		15R	6H	0.5895 (Na)		300	Thibault B
	n_o	2.6467	2.648				
	n_e	2.690	2.689				
Dispersion	dn_o/d_λ	0.0918		λ= 4360-6200Å	20-460°C	Ramdas	
	dn_e/d_λ	0.1028					
Temperature Coeff. $(1/n)(dn/dT)$ (o,ε)		$3.5\mathrm{x}10^{-5}$/°K		λ= 4710-6680Å	20-460°C	Ramdas	
Laser Wavelength		4560	Å		300	Griffiths et al.	
Linewidth		<5	Å		300	Griffiths et al.	
Threshold Current Density		120	A/cm^{2}		300	Griffiths et al.	

ADAMSKY, R.F. and K.M. MERZ. Synthesis and Crystallography of the Wurtzite Form of Silicon Carbide. Z. FUER KRISTALLOGRAPHIE, v. 111, 1959. p. 350-361.

ARLT, G. and G.R. SCHODDER. Some Elastic Constants of Silicon Carbide. ACOUSTICAL SOC. OF AMERICA, J., v. 37, no. 2, Feb. 1965. p. 384-386.

BARRETT, D.L. Evaluation of Trace Impurities in the Preparation of High-Purity Silicon Carbide. ELECTROCHEM. SOC., J., v. 113, no. 11, Nov. 1966. p. 1215-1218.

BARRETT, D.L. and R.B. CAMPBELL. Electron Mobility Measurements in Silicon Carbide Polytypes. J. OF APPLIED PHYS., v. 38, no. 7, Jan. 1967. p. 53-55.

BELLE, M.L., et al. Preparation and Optical Properties of Cubic beta-Silicon Carbide. SOVIET PHYS.-SEMICONDUCTORS, v. 1, no. 3, Sept. 1967. p. 315-319.

BOSCH, G. On the Thermal Conductivity of Silicon Carbide. PHILIPS RES. REPTS., v. 16, no. 5, Oct. 1961. p. 455-461.

UNION CARBIDE CORP. PARMA RES. LAB. Thermoelectric Materials. By: BRECKENRIDGE, R.G. Bi-Monthly Pr. no. 8, Mar. 28-May 28, 1960. Contract no. NObs-77066. June 15, 1960. ASTIA AD-245 092.

BRODSKY, M.H. and R.S. TITLE. Electron Spin Resonance in Amorphous Silicon, Germanium, and Silicon Carbide. PHYS. REV. LETTERS, v. 23, no. 11, Sept. 1969. p. 581-585.

ROYAL AIRCRAFT ESTABLISHMENT. Silicon Carbide - A Review. By: BROWN, A.R.G. Tech. Note no. Met/Phy 325. ASTIA AD-249 685.

BROWN BOVERI RESEARCH CENTER. Solid State Physics. SWISS TECHNICS, no. 2, 1970. p. 49.

BUKKE, E.E. et al. Optical Properties of Acceptors in Silicon Carbide Crystals. SOVIET PHYS.-SEMICONDUCTORS, v. 1, no. 9, Sept. 1967. p. 1164-1167.

CARNAHAN, R.D. Elastic Properties of Silicon Carbide. AMERICAN CERAM. SOC., J., v. 51, no. 4, Apr. 1968. p. 223-224.

CHANG, H. et al. Use of Silicon Carbide in High Temperature Transistors. CONF. ON SILICON CARBIDE, BOSTON, 1959. Ed. by: O'CONNOR, J.R. and J. SMILTENS. N.Y. Pergamon Press, 1960. p. 496-507.

CHOYKE, W.J. et al. Optical Properties of Cubic Silicon Carbide. Luminescence of Nitrogen-Exciton Complexes and Interband Absorption. PHYS. REV., v. 133, no. 4A, Feb. 17, 1964. p. A1163-A1166. (A)

CHOYKE, W.J. et al. Optical Properties of 4H Silicon Carbide - Absorption and Luminescence. INT. CONF. ON SEMICONDUCTOR PHYS., PROC., 7th, Paris, 1964, v. 1. Ed. by: HULIN, M. N.Y., Acad. Press, 1964. p. 751-758. (B)

CHOYKE, W.J. et al. Exciton Complexes and Donor Sites in 33R Silicon Carbide. PHYS. REV., v. 139, no. 4A, Aug. 19, 1965. p. A1262-A1274. (C)

CHOYKE, W.J. and L. PATRICK. Higher Absorption Edges in Cubic Silicon Carbide. PHYS. REV., v. 187, no. 3, Nov. 15, 1969. p. 1041-1043. (A)

CHOYKE, W.J. and L. PATRICK. Exciton Recombination Radiation and Phonon Spectrum of 6H Silicon Carbide. PHYS. REV., v. 127, no. 6, Sept. 15, 1962. p. 1868-1877. (B)

CHOYKE, W.J. and L. PATRICK. Higher Absorption Edges in 6H Silicon Carbide. PHYS. REV., v. 172, no. 3, Aug. 15, 1968. p. 769-772. (C)

DALVEN, R. Temperature Coefficient of the Energy Gap of beta-Silicon Carbide. J. OF PHYS. AND CHEM. OF SOLIDS, v. 26, no. 2, Feb. 1965. p. 439-441.

DAS, D. Magnetic Studies on alpha-Silicon Carbide Crystals. INDIAN J. OF PHYS., v. 41, no. 7, July 1967. p. 525-532.

DE MESQUITA, A.H. G. Refinement of the Crystal Structure of SiC Type 6H. ACTA CRYSTALLOGRAPHICA, v. 23, 1967. p. 610-617.

DILLON, J.A., Jr. et al. Some Surface Properties of Silicon-Carbide Crystals. J. OF APPLIED PHYS., v. 30, no. 5, May 1959. p. 675-679.

DONNAY, J.D.H. (Ed.) Crystal Data. Determinative Tables. 2nd Ed. AMERICAN CRYSTALLOGRAPHY ASSOC., 1963.

DUBROVSKII, G.B. Emissivity of Semiconducting Silicon Carbide at High Temperatures. SOVIET PHYS.-SEMICONDUCTORS, v. 3, no. 10, Apr. 1970. p. 1290-1293.

MARTIN CO., MARTIN-MARIETTA CORP., ORLANDO, FLA. Infrared Signature Characteristics. By: DURAND, J.L. and C.K. HOUSTON. ATL-TR-66-8. Contract no. AF 08 635 5087, Jan. 1966. 174 p. AD 478 597.

ELLIS, B. and T.S. MOSS. The Conduction Bands in 6H and 15R Silicon Carbide. I. Hall Effect and Infrared Faraday Rotation Measurements. ROYAL SOC. OF LONDON, PROC., v. 299, Ser. A, no. 1458, July 11, 1967. p. 383-392.

GHOSHTAGORE, R.N. and R.L. COBLE, Self-Diffusion in Silicon Carbide. PHYS. REV., v. 143, no. 2, Mar. 1966. p. 623-626.

GMELINS HANDBUCH DER ANORGANISCHEN CHEMIE; achte völlig neu bearbeitete Auflage. Silicium. Teil B. Weinheim, Verlag Chemie, GmbH, 1959.

GORBAN, I.S. et al. Energy and Kinetic Parameters of Nitrogen Impurity in Silicon Carbide Crystals. SOVIET PHYS. SOLID STATE, v. 8, no. 11, May 1967. p. 2746-2747. (B)

GORBAN, I.S. et al. Energy Spectrum of Acceptors Acting as Radiative Recombination Centres in SiC. SOVIET PHYS. SEMICONDUCTORS, v. 1, no. 4, Oct. 1967. p. 514-515. (A)

GORBAN, I.S. and A.P. KROKHMAL. Absorption Spectrum of Four-Particle Nitrogen-Exciton Complexes in Silicon Carbide. SOVIET PHYS. SOLID STATE, v. 12, no. 3, Sept. 1970. p. 699-700.

GORYUNOVA, N.A. The Chemistry of Diamond-Like Semiconductors. Ed. by: ANDERSON, J.C. The MIT Press, Mass. Inst. of Tech., Cambridge, Mass., 1963. 236 p.

GRIFFITHS, L.B. Nature of Rectifying Junctions in alpha-Silicon Carbide. J. OF APPL. PHYS., v. 36, no. 2, Feb. 1965. p. 571-575.

GRIFFITHS, L.B. et al. Silicon Carbide Diode Laser. IEEE, PROC., v. 51, no. 10, Oct. 1963. p. 1374-1376.

HAGEN, S.H. and C.J. KAPTEYNS. The Ionization Energy of Nitrogen Donors in 6H and 15R Silicon Carbide. PHILIPS RES. REPTS., v. 25, no. 1, Feb. 1970. p. 1-7.

HAMILTON, D.R. et al. Photoluminescence of Nitrogen-Exciton Complexes in 6H Silicon Carbide. PHYS. REV., v. 131, no. 1, July 1, 1963. p. 127-133. (A)

HAMILTON, D.R. et al. Optical Properties of 21R Silicon Carbide - Absorption and Luminescence. PHYS. REV., v. 138, no. 5A, May 31, 1965. p. A1472-A1476. (B)

HOFMAN, D. et al. The Dielectric Constant of Silicon Carbide. PHYSICA, v. 23, 1967. p. 236.

KALNIN, A.A. et al. Photoluminescence of Silicon Carbide with Beryllium Impurity. SOVIET PHYS.-SOLID STATE, v. 8, no. 10, Apr. 1967. p. 2381-2383.

KAN, Kh.S. and B.S. KULVARSKAYA. On the Thermionic Emission Mechanism of Refractory-Metal Carbide Cathodes. ACAD. OF SCI., USSR, BULL., PHYS. SER., v. 33, no. 3, 1969. p. 409-415.

KERN, E.L. et al. Thermal Properties of beta Silicon Carbide from 20 to 2000°C. MAT. RES. BULL., v. 4, p. S25-S32. Proc. of Int. Conf. on Silicon Carbide, University Park, Pa., Oct. 1968. Pergamon Press, N.Y. 1969.

KHOLYUANOV, G.F. The Roles of Boron, Nitrogen, and Gallium in the Electroluminescence of Silicon Carbide p-n Junctions. SOVIET PHYS.-SOLID STATE, v. 7, no. 11, May 1966. p. 2620-2624.

KROKO, L.J. and A.G. MILNES. Diffusion of Nitrogen into Silicon Carbide Single Crystals Doped with Aluminum. SOLID STATE ELECTRONICS, v. 9, no. 12, Dec. 1966. p. 1125-1134.

KULVARSKAYA, B.S. et al. Thermionic Emission of Certain Refractory Materials and Possible Use as Cathodes in Gaseous Devices. SOVIET PHYS. TECH. PHYS., v. 14, no. 1, July 1969. p. 122-128.

LANDOLT-BÖRNSTEIN. Numerical Data and Functional Relationships in Science and Technology. New Series. Ed. by: K-H. HELLWEGE. Group III, Crystal and Solid State Phys., v. 2, SPRINGER-VERLAG, Berlin, Germany, 1969. p. 61.

LEZHENIN, F.F. and G.G. GNESIN. Thermal Conductivity of Silicon Carbide at High Temperatures. SOVIET POWDER METALL. AND METAL CERAM., v. 7, no. 2, Feb. 1967. p. 114-116.

LISITSA, M.P. et al. Polytypism and Low Temperature Photoluminescence of Silicon Carbide Single Crystals. SOVIET PHYS.-SOLID STATE, v. 12, no. 4, Oct. 1970. p. 1014-1015.

LOMAKINA, G.A. and Yu. A. VODAKOV. Effect of Phonon Drag in alpha-Silicon Carbide Crystals. SOVIET PHYS.-SOLID STATE, v. 4, no. 3, Sept. 1962. p. 603-604.

MASLAKOVETS, Yu.P. et al. Diffusion of Beryllium in Silicon Carbide. SOVIET PHYS.-SOLID STATE, v. 10, no. 3, Sept. 1968. p. 634-638.

MOKHOV, E.N. et al. Diffusion of Aluminum in Silicon Carbide.SOVIET PHYS.-SOLID STATE, v. 11, no. 2, Aug. 1969. p. 415-416.

NELSON, W.E. et al. Growth and Properties of beta-Silicon Carbide Single Crystals. J. OF APPLIED PHYS., v. 37, no. 1, Jan. 1966. p. 333-336.

OZAROW, V. and R.E. HYSELL. Space-Charge-Limited Currents in Single Crystal Silicon Carbide. J. OF APPLIED PHYS., v. 33, no. 10, Oct. 1962. p. 3013-3015.

PATRICK, L. et al. Optical Properties of 15R Silicon Carbide. Luminescence of Nitrogen-Exciton Complexes and Interband Absorption. PHYS. REV., v. 132, no. 5, Dec. 1, 1963. p. 2023-2031. (A)

PATRICK, L. et al. Growth, Luminescence, Selection Rules, and Lattice Sums of Silicon Carbide with Wurtzite Structure. PHYS. REV., v. 143, no. 2, Mar. 1966. p. 526-536. (B)

PATRICK, L. High Electron Mobility of Cubic Silicon Carbide. J. OF APPL. PHYS., v. 37, no. 13, Dec. 1966. p. 4911-4913.

PATRICK, L. and W.J. CHOYKE. Static Dielectric Constant of Silicon Carbide. PHYS. REV. B, v. 2, no. 6, Sept. 1970. p. 2255-2256.

PHILIPP, H.R. and E.A. TAFT. Intrinsic Optical Absorption in Single Crystal Silicon Carbide. CONF. ON SILICON CARBIDE, BOSTON, 1959. Ed. by: O'CONNOR, J.R. and J. SMILTENS. Pergamon Press, N.Y. 1960. p. 366-371.

PICHUGIN, I.G. and A.N. PIKHTIN. Impurity Absorption in Silicon Carbide Crystals with Boron during Crystal Growth. SOVIET PHYS.-SOLID STATE, v. 8, no. 2, Feb. 1966. p. 465-466.

MARTIN CO., MARTIN-MARIETTA CORP., ORLANDO, FLA. Research on Optical Properties of Single Crystals of beta Phase Silicon Carbide. By: PICKAR, P.B. et al. Summary Tech. Rept. no OR8394, May 5, 1965-June 5, 1966. Contract no. DA-28-017-AMC-2002(A). Oct. 1966. 85 p. AD 641 198.

PIKHTIN, A.N. and D.A. YAS'KOV. Fundamental Absorption Edge in 6H Silicon Carbide. SOVIET PHYS.-SOLID STATE, v. 12, no. 6, Dec. 1970. p. 1267-1272.

PONG, W. and K. FUJITA. Photoemission from Silicon-Carbon Crystals in the Vacuum Ultraviolet. J. OF APPLIED PHYS., v. 37, no. 1, Jan. 1966. p. 445.

RAMDAS, A.K. Thermo-Optic Behaviour of Silicon Carbide. INDIAN ACAD. OF SCI., PROC., v. 34a, 1951. p. 136-140.

ROSENGREEN, A. Electronic Properties of n-Type beta Silicon Carbide Crystals Grown from Solution. MAT. RES. BULL., v. 4, p. S355-S364. Proc. of Int. Conf. on Silicon Carbide, University Park, Pa., Oct. 1968. Pergamon Press, N.Y. 1969.

SCACE, R.I. and G.A. SLACK. The Silicon Carbide and Germanium Carbide Phase Diagrams. CONF. ON SILICON CARBIDE, BOSTON, 1959. Ed. by: O'CONNOR, J.R. and J. SMILTENS. Pergamon Press, N.Y., 1960. p. 24-30.

SHAFFER, P.T.B. A Review of the Structure of Silicon Carbide. ACTA CRYST., v. B25, 1969. p. 477-488. (A)

SHAFFER, P.T.B. Effect of Crystal Orientation on Hardness of Silicon Carbide. AMERICAN CERAM. SOC., J., v. 47, no. 9, Sept. 1964. p. 466. (B)

SHAFFER, P.T.B. Effect of Crystal Orientation on Hardness of beta Silicon Carbide. AMERICAN CERAM. SOC., J., v. 48, no. 11, Nov. 1969. p. 602. (C)

PLENUM PRESS HANDBOOK OF HIGH TEMPERATURE MATERIALS. No. 1., Materials Index. By: SHAFFER, P.T.B. Plenum Press, N.Y., 1964. p. 107-111. (D)

SHAFFER, P.T.B. and R.G. NAUM. Refractive Index and Dispersion of beta Silicon Carbide. OPTICAL SOC. OF AMERICA, J., v. 59, no. 11, Nov. 1969. p. 1498.

STANFORD RES. INST., MENLO PK., CALIF. Development of Manufacturing Methods for Growing Large beta-Silicon Carbide Single Crystals. IR-8-222 VI. By: SILVA, W.J. et al. Contract no. F336 15-62-C-1328. Dec. 1967. 52 p.

SLACK, G.A. Thermal Conductivity of Pure and Impure Silicon Carbide and Diamond. J. OF APPLIED PHYS., v. 35, no. 12, Dec. 1964. p. 3460-3466.

SLACK, G.A. and R.I. SCACE. Nitrogen Incorporation in Silicon Carbide. J. OF CHEM. PHYS., v. 42, no. 2, Jan. 1965. p. 805-806.

SPITZER, W.G. Infrared Properties of Hexagonal Silicon Carbide. PHYS. REV., v. 113, no. 1, Jan. 1, 1959. p. 127-132.

TAYLOR, A. and R.M. JONES. The Crystal Structure and Thermal Expansion of Cubic and Hexagonal Silicon Carbide. CONF. ON SILICON CARBIDE, BOSTON, 1959. Ed. by: O'CONNOR, J.R. and J. SMILTENS. Pergamon Press, N.Y. 1960. p. 147-154.

THIBAULT, N.W. alpha-Silicon Carbide, Type 51R. AMERICAN MINERALOGIST, v. 33, 1949. p. 588-599. (A)

THIBAULT, N.W. Morphological and Structural Crystallography and Optical Properties of Silicon Carbide. AMERICAN MINERALOGIST, v. 39, no. 9-10, Sept.-Oct. 1944. p. 327-362. (B)

TOLPYGO, K.B. Optical, Elastic and Piezoelectric Properties of Ionic and Valence Crystals with the Zinc-Sulfide Type Lattice. SOVIET PHYS.-SOLID STATE, v. 2, no. 10, Apr. 1961. p. 2367-2376.

VAKHNER, Kh. and Yu.M. TAIROV. Luminescence of Silicon Carbide Doped with Scandium. SOVIET PHYS.-SOLID STATE, v. 11, no. 9, Mar. 1970. p. 1972-1974. (B)

VAKHNER, Kh. and Yu.M. TAIROV. Polytypism of Scandium-doped Silicon Carbide Crystals Grown from Solution. SOVIET PHYS.-SOLID STATE, v. 12, no. 5, Nov. 1970. p. 1213-1214. (A)

VAKULENKO, O.V. and O.A. GOVOROVA. Optical Absorption of 6H alpha-Silicon Carbide Crystals Containing Nitrogen. SOVIET PHYS.-SOLID STATE, v. 12, no. 4, Dec. 1970. p. 1478-1479.

PHILIPS RES. REPTS. Mobility of Charge Carriers in Silicon Carbide. By: VAN DAAL, H.J. Rept. Suppl. no. 3. 1965. N65-30608.

VAN DAAL, H.J. et al. On the Electronic Conduction of alpha-Silicon Carbide Crystals between 300 and 1500°K. J. OF PHYS. AND CHEM. OF SOLIDS, v. 24, Jan. 1963. p. 109-127.

VAN RYNEVELD, W.P. and J.H.N. LOUBSER. Electron Spin Resonance in Electron-Irradiated n-type Silicon Carbide. BRITISH J. OF APPL. PHYS., v. 17, no. 10, 1966. p. 1277-1283.

VODAKOV, Y.A. et al. Diffusion of Boron and Aluminium in n-type Silicon Carbide. SOVIET PHYS.-SOLID STATE, v. 8, no. 4, Oct. 1966. p. 1040-1041.

WYCKOFF, R.W.G. Crystal Structures. N.Y. Wiley and Sons, v. 1. p. 114.

YU, HSIAO-TUNG et al. Wu Li Hsuech Pao, v. 22, no. 9, 1966. p. 976-981.

ZANMARCHI, G. Luminescence of the Aluminium Centre in Cubic Silicon Carbide. Dependence of the Recombination Rates on the Intensity of the Light Excitation. J. OF PHYS. AND CHEM. OF SOLIDS, v. 29, no. 10, Oct. 1968. p. 1727-1736. (A)

ZANMARCHI, G. Absorption of Light Near the Band Edge in alpha-Silicon Carbide. PHILIPS RES. REPTS., v. 20, no. 3, June 1965. p. 253-268. (B)

ZHUKOVA, I.I. et al. Investigation of the Energy Structure of Silicon Carbide and Silicon Nitride by Ultrasoft X-Ray Spectroscopy. SOVIET PHYS.-SOLID STATE, v. 10, no. 5, Nov. 1968. p. 1097-1103.

ZHUMAKULOV, U. and N.A. SMIRNOVA. Optical Reflection Spectra of Gallium Phosphide, Silicon Carbide, and Boron. SOVIET PHYS. SEMICONDUCTORS, v. 3, no. 5, Nov. 1969. p. 654.

ZIOMEK, J.S. and P.B. PICKAR. Optical Studies of beta-Silicon Carbide. PHYS. STATUS SOLIDI, v. 21, no. 1, May 1967. p. 271-278.

MIX
Papier aus verantwortungsvollen Quellen
Paper from responsible sources
FSC® C105338

If you have any concerns about our products,
you can contact us on
ProductSafety@springernature.com

In case Publisher is established outside the EU,
the EU authorized representative is:
Springer Nature Customer Service Center GmbH
Europaplatz 3, 69115 Heidelberg, Germany

Printed by Libri Plureos GmbH
in Hamburg, Germany